Whole Building Life Cycle Assessment

Reference Building Structure and Strategies

WBLCA Guide Special Project Working Group

Edited by Frances Yang, S.E.

Sponsored by
Sustainability Committee of the Structural Engineering Institute
of the American Society of Civil Engineers

ASCE

SEI ASCE STRUCTURAL ENGINEERING INSTITUTE

Published by the American Society of Civil Engineers

Library of Congress Cataloging-in-Publication Data

Names: Yang, Frances, editor.
Title: Whole building life cycle assessment : reference building structure and strategies / edited by Frances Yang.
Description: Reston, Virginia : American Society of Civil Engineers, [2018] | "Sponsored by Sustainability Committee of the Structural Engineering Institute of the American Society of Civil Engineers, WBLCA Guide Special Project Working Group." | Includes bibliographical references and index.
Identifiers: LCCN 2018016350 | ISBN 9780784415054 (soft cover : alk. paper) | ISBN 9780784481448 (pdf)
Subjects: LCSH: Sustainable construction. | Structural engineering. | Building materials—Service life. | Ecological risk assessment. | Service life (Engineering) | Example.
Classification: LCC TH880 .W47 2018 | DDC 690.028/6—dc23
LC record available at https://lccn.loc.gov/2018016350

Published by American Society of Civil Engineers
1801 Alexander Bell Drive
Reston, Virginia 20191-4382
www.asce.org/bookstore | ascelibrary.org

***Errata:* Errata, if any, can be found at https://doi.org/10.1061/9780784415054.**

ISBN 978-0-7844-1505-4 (print)
ISBN 978-0-7844-8144-8 (PDF)
Manufactured in the United States of America.

24 23 22 21 20 19 18 1 2 3 4 5

About the Cover

Capitol Tower, developed and constructed by Skanska and designed by Gensler, with Walter P. Moore as structural engineer, participated in the LEED v4 Beta program and was one of the first projects in the nation to pursue the LEED v4 Whole Building Life-Cycle Assessment (WBLCA) credit. The 775,000 sq-ft office building in downtown Houston achieved LEED v4 Platinum Pre-Certification and will be the most sustainable office building in the city. The structure topped out in April 2018. The WBLCA informed the design strategy and focused on concrete optimization of the 35-story concrete frame and foundation. Both achieved the LEED v4 thresholds and saved more than 8,000 tons of carbon emissions, approximately equivalent to the emissions of driving 15 million miles.

Portola Valley Town Center. Siegel & Strain Architects and structural engineers Forell/Elsesser employed salvaged glulam beams, as well as whole trunks of cleared trees for timber columns, and repurposed many elements from disassembly of the previous building on-site into various nonstructural elements. They also used a high slag mix to lower the cement in the concrete. This was one of the first US projects to conduct a full whole-building LCA, which resulted in a 125-ton carbon savings, amounting to about 32% embodied carbon savings due to these strategies.

Contents

Acknowledgments

This document is the result of a collaborative effort by a subgroup of the ASCE/SEI Sustainability Committee, made possible through the Special Projects grant of ASCE. Each author contributed practical experience and a different point of view in the buildings industry to ensure the guidance represents the most contemporary, balanced information on the practice of comparative whole building life cycle assessment.

Primary Authors

Dirk Kestner
Lionel Lemay
Adam Slivers
Megan Stringer
Mark Webster
Lauren Wingo
Erika Winters-Downey
Frances Yang *(Project Lead)*

The authors gratefully acknowledge the following guest reviewers, hand-selected for their expertise and deep involvement in the growth of this field within the buildings industry.

Matthew Bowick, Athena Sustainable Materials Institute

Sarah Buffaloe, WSP (formerly with US Green Building Council)

Erica Fischer, Oregon State University

Bradley Guy, Catholic University

Jordan Palmeri, Oregon Department of Environmental Quality

Nick Santero, Apple Inc. (formerly with thinkstep, a.k.a. PE International)

David Shook, SOM

Kathrina Simonen, University of Washington

Kyle Steuck, Degenkolb

Wes Sullens, US Green Building Council (formerly at StopWaste.org)

Wayne Trusty, Wayne B. Trusty & Associates Ltd.

Martha VanGeem, self-employed

In addition, we wish to thank the following members of the Sustainability Committee for their constructive review:

Laura Dolak, WSP

Emily Lorenz, SevGen Consulting

Christopher Horiuchi, SOM

David DeLong, IC Design Services

The authors would also like to especially recognize Kelly Roberts of Walter P. Moore for her editorial help on the document and Loralie Froman for professional copyediting.

We hope that this guide will be continuously updated to reflect the best practices and latest advances in comparative whole building life cycle assessment as it pertains to structural engineers.

Introduction

This technical report serves as a guide for use by the project team to define and model the structural system within the "reference building design," as required by green building standards and rating systems when performing the whole building life cycle assessment (WBLCA). This guide supports the prescribed use of WBLCA by these systems to show how project teams have made improvements to reduce environmental impacts through the project team's design and construction choices.

This report does not discuss the process of conducting a WBLCA or compare the standards and rating systems prescribing use of WBLCA. The scope of this document is limited to

- Definition of the "reference building design,"
- Design and construction choices within the influence of the structural engineer, and
- Discussion of the approach to constructing the reference building design influenced by choice of the structural system and materials.

This technical guide assumes that the user possesses basic knowledge on life cycle assessment for whole buildings If the user needs to establish these fundamentals, existing WBLCA standards and guidelines and Appendix C of this guide provide a good introduction for novice users. The recommended literature should be studied in conjunction with employing at least one of the WBLCA tools; many are listed in "Appendix B: LCA Tools."

The report is organized in two parts. The first part sets out terms of reference, including definitions, reference building options, and general considerations to keep in mind when defining the reference building design. The second part describes specific strategies for the reference building a structural engineer may choose to employ for reducing life cycle impacts on a project. These strategies are not inclusive of all possible strategies that may prove effective in reducing the life cycle environmental impacts of buildings. Rather, they are the strategies the authors deemed most common or effective and also most relevant to the role of the structural engineer.

WBLCA is a relatively new application of life cycle assessment (LCA). Because LCA has primarily been used for mass-produced products, applying the principals and method to buildings—which are for the most part one-off creations—has some challenges. Although some of these challenges are addressed here, the authors acknowledge that the guidance offered is preliminary and will continue to improve as the field of WBLCA grows and matures with its users.

The authors of this technical guide also recognize that environmental performance is only one of the factors driving the decisions made on a project. Design and construction decisions will continue to demand a balanced consideration of oftentimes competing priorities, including cost, constructability, durability, operational performance, and the assumed values of future occupants. However, green building standards and rating systems are a growing driver in these decisions, as is the adoption of comparative WBLCA. Yet, within the requirements for conducting LCA-based comparisons, there is little detailed guidance regarding the structural system and materials.

A concern arose within the Structural Engineering Institute's Sustainability Committee that the current lack of direction could allow the project team to construct an unrealistic reference building that showed a false improvement in the proposed design. The committee felt this situation could fail to encourage the meaningful contribution structural engineers might make toward reducing real environmental impacts of our building structures. The authors, therefore, have produced this guide to bridge the gap and ensure a nominal level of integrity in comparing structural designs using WBLCA. Ultimately, this guide aims to encourage structural engineers to take a more active and meaningful role in the practice of green building and life-cycle assessment. The authors are structural engineers writing primarily for structural engineers, but other building professionals will also find value in this report.

PART I

Terms of Reference

Glossary

Most of the definitions provided are based on definitions from EN 15978 or 15804.

Baseline Building
: Synonymous with "reference building." The USGBC's LEED rating system uses the term "baseline building," whereas this document, ASTM E2921, and other guidance use "reference building."

Building Code
: A building code is a set of rules that specifies construction standards for buildings and non-building structures as well as codes regulating the design and construction of structures, where adopted into law by a legal entity, such as a federal, state, county, or municipal government or agency.

Environmental Product Declaration
: An environmental product declaration (EPD) is a standardized way of reporting the environmental impacts of a product, process, or service from a life cycle assessment.

Functional Equivalence
: The condition that occurs when the baseline building and proposed building meet the same functional and/or technical requirements (e.g., meeting the same building type, pattern of use, service life, and so on).

Green Building Programs
: A rating system, standard, or code in which the technical requirements or guidelines provide for evaluating the sustainability of a building and its components.

Impact Indicator
: A measure used to represent an environmental area of concern deemed relevant in life cycle assessment (LCA), which is quantified and subsequently assigned a value during the life cycle inventory analysis phase of LCA. See the Impact Indicators section of the General Considerations chapter for further detail.

Life Cycle Assessment (LCA)
: Life cycle assessment is a technique for assessing the environmental impacts associated with a product, process, or service that takes into account the impacts from raw material extraction processing, manufacture, distribution, use, or repair and maintenance through its disposal or recycling.

Life Cycle Inventory (LCI)

Life cycle inventory data quantify the material flows, energy flows, and environmental impacts of a given product. LCI data are used as input to a life cycle assessment.

Material

The substance or products used for the structure or building such as concrete, steel, timber, or masonry.

Performance-Based Design

Performance-based design is a formal process for the design and retrofit of buildings intended to achieve target performance objectives corresponding to various levels of hazards expected at the site. Performance objectives relate to acceptable impacts in terms of habitability, continuity of operations, and damage.

Product Category Rule

A product category rule is a set of specific rules, requirements, and guidelines for developing environmental product declarations.

Project Team

The members of the team, including the owner, who design and construct a building project.

Proposed Building

The proposed design for which the project team is seeking compliance with green building standards or rating systems and which requires a comparative WBLCA; synonymous with "as-design building" or "actual design."

Rating System

Voluntary system of classification according to quality or amount to which industry technical guidelines are applied.

Recycling

A recovery operation through which waste materials are reprocessed into products, materials, or substances.

Reference Building

The scenario building used for comparison of environmental impacts to the proposed building through the performance of an LCA. Following are three starting options for defining the reference building, as described further in the chapter titled Reference Building Options.

Existing Building: The starting option for the reference building that is the design of an existing building or buildings.

Building Archetype: The starting option for the reference building that is a designed-but-unconstructed building that represents a "typical" building within a defined set of parameters.

Early Design: The starting option for the reference building that is an earlier design for the building under study.

Reuse

Any operation through which products or components are used again for the same purpose or are used for other purposes without reprocessing.

Service Life

The estimated length of time a building or assembled system will be in use after installation, typically defined by the green building program.

Specification

The characteristics of a material, product, or system defined in the construction documents that dictate specific attributes of the material, product, or system to meet project criteria.

Standard

Technical requirements established by authority, custom, or general consent as a model or example that describes a level of quality, achievement, and so on that is considered acceptable or desirable. Standards are voluntary unless referenced by building code.

Structural System

The type of structural framing used to support loads, differentiated by trades and design methodology. The components of the primary structural frame as commonly defined as the foundations, bearing walls, shear walls, columns, beams, slabs, and lateral bracing required to maintain the stability of the final structure as a whole.

Examples

Flat plate and composite concrete-filled metal deck (for slab systems); precast concrete, cast-in-place concrete, prestressed concrete, steel frame, timber stick frame, heavy mass timber, and masonry (for gravity frames); braced frame, moment frame, and shear walls (for lateral systems); mat, pile, and spread footings (for foundations); and various retaining walls systems, tension membrane fabric systems, light-gauge metal building systems, and inflatable structures.

Whole Building Life Cycle Assessment (WBLCA)

A compilation and evaluation of the inputs, outputs, and potential environmental impacts of a whole building throughout its life cycle.

General Considerations

EXISTING WBLCA STANDARDS AND GUIDELINES

Readers are encouraged to refer to the following two documents for important, overarching guidance in defining the reference building.

ASTM E2921. *Standard Practice for Minimum Criteria for Comparing Whole Building Life Cycle Assessments for Use with Building Codes and Rating Systems* (ASTM 2016).

This standard sets out criteria when LCA is undertaken at the whole building level to compare a final whole building design to a reference building design. The limitations set out in ASTM E2921 constrain differences on parameters influencing the operational energy performance, the reference service life of products, and how product replacement needs to be considered for the whole building LCA. It also stipulates that the user needs to use the same software tool for both buildings and to follow ISO 14040 and ISO 14044. While the standard marks a milestone in setting criteria and boundaries on the comparison, it is a concise document (mostly because it points to ISO 14040 and 14044) and does not detail how structural systems and materials may or may not differ between the reference and proposed designs. Thus, this document supplements E2921 with the intent that the project team should follow ASTM E2921 in addition to the limitations listed in the Reference Building Options chapter and the chapters within Part II: Strategies, in this book.

Athena Guide to Whole-Building LCA in Green Building Programs (Athena WBLCA Guide).

To date, this guidance document offers the most comprehensive comparison of WBLCA requirements across the most popular green building rating systems and standards in the United States and Canada. Readers should reference Table 2 in the Athena WBLCA Guide for various rating system requirements.

While this guide attempts to remain impartial with respect to the various standards and systems, readers should follow the requirements of the green building standard or rating system prescribed for the project.

Other relevant LCA standards and guidelines:

ISO 14040. *Environmental management—Life cycle assessment—Principles and framework* (ISO 2006).

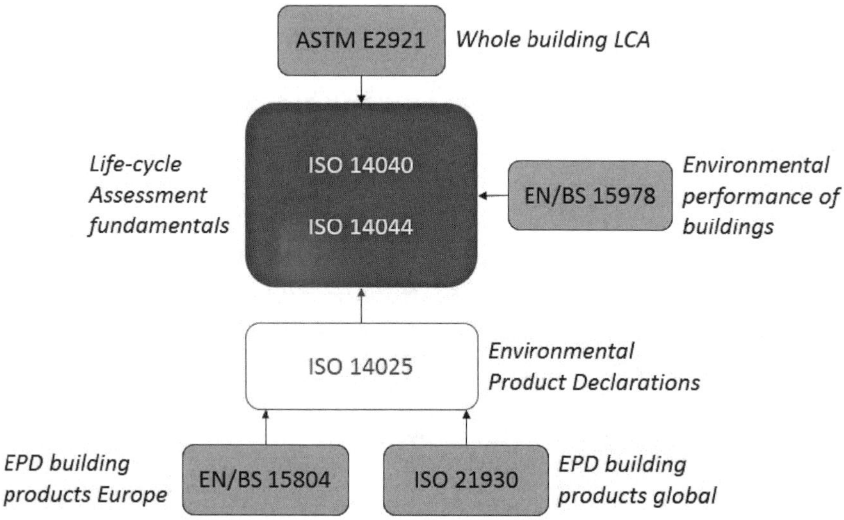

Figure 1: Relationship of relevant LCA Standards and Guidelines
Source: Adapted with permission from Simonen (2014).

ISO 14044. *Environmental management—Life cycle assessment—Requirements and guidelines* (ISO 2006).
ISO 21930. *Sustainability in building construction—Environmental declaration of building products* (ISO 2007).
EN/BS 15804. *Sustainability of construction works, Environmental product declarations: Core rules for the product category of construction products* (EN/BS 2012).
EN/BS 15978. *Sustainability of construction works: Assessment of environmental performance of buildings, Calculation method* (EN/BS 2011).

Green Building Programs: Rating Systems, Standards, and Codes

The following rating systems, standards, and codes contain mandatory or voluntary requirements for conducting a WBLCA. Readers should follow those that are relevant to their project. This document references some of them to clarify concepts but does not attempt to exhaustively compare or explain the requirements set forth in these rating systems, standards, and codes.

Rating Systems	Standards	Codes
Green Globes	ASHRAE 189.1-14 (2014)	CalGreen (2013)
LEED v4 (mandatory 2016)	ICC 700 (2012)	IgCC (2015)

OBJECT OF ASSESSMENT

Defining the object of assessment of an LCA is the project team's first step and determines the elements of the building to be included in the assessment. Generally, some building systems are excluded from the LCA (e.g., mechanical, plumbing, electrical, fixtures, etc.), as well as site work and landscape. Whichever systems are excluded in the reference building should also be excluded in the proposed building.

Notes

At a minimum, most WBLCAs should include structural and envelope materials, as these systems usually account for the majority of the material-embodied impacts (Arup 2010; Hsu 2010; Ochsendorf et al. 2011). The green building program guiding the WBLCA generally specifies which systems should be included. Structural materials include primary structural components comprising the foundations, bearing walls, shear walls, columns, beams, slabs, and lateral bracing required to maintain the stability of the final structure as a whole.

The object of assessment as prescribed by a few green building programs:

- ASTM E2921 requires the inclusion of structure, envelope, conduit, ductwork, piping and wiring, and interior walls, floors, and ceilings, but not electrical and mechanical equipment, plumbing fixtures, fire safety systems, or conveying systems. Finishes should be included to the extent data are available.
- LEED v4 requires the inclusion of complete building envelope and structural elements, including the material components of footings and foundations, structural wall assemblies (from cladding to interior finishes), structural floors and ceilings (not including finishes), roof assemblies, and parking structures.
- Tables 2 through 6 of the Athena WBLCA Guide show which systems need to be included for the various green building programs.

The design team should coordinate across all relevant disciplines to fully define the object of assessment.

SYSTEM BOUNDARY

The system boundary defines the life cycle stages included in the object of assessment, which should be identical between the reference building and the proposed building. Life cycle stages may be defined using the format presented in BS/EN 15978 (which utilizes the definitions in BS/EN 15804). Refer to the guiding WBLCA green building program for the required system boundaries, as they may differ.

Notes

The green building program guiding the WBLCA typically specifies which life cycle stages should be included in the analysis. Both ASTM E2921 and LEED v4 require cradle-to-grave assessments of material impacts (see Table 1: Life Cycle Stages and Corresponding LCA Strategies, for the names of the life cycle stage modules). The cradle-to-gate (A1–A3) and cradle-to-grave (A1–C4) or (A1–D) modules are the most common boundaries for products, while cradle-to-grave is typically required for whole building LCAs. The inclusion of Module D is optional in cradle-to-grave assessments. It is often allowable for the project team to exclude specific modules, such as B6 (operational energy use) and B7 (operational water use), in WBLCA used for green building programs. See the Athena WBLCA Guide, Table 4, for which phases may or may not be required by the applicable green building program.

Table 1 shows the life cycle phases in modules, using the format defined in BS/EN 15804, and the corresponding strategies found in Part II: Strategies. In all cases, the same life cycle phases should be included for both the reference and project buildings. (i.e., the extent of life cycle phases should match), although the quantity and environmental performance of the materials may differ.

Material end-of-life recycling or reuse beyond the system boundary (Module D) may be included in the LCA but should be reported separately.

Notes

The inclusion of Module D in the LCA is considered controversial because material disposition many decades in the future is often uncertain.

A structural engineer should consider the structure's impacts during the use phase with respect to the following:

- Energy (energy use may be influenced by the choice of structural system). For discussion of coordination with energy consultants, see the strategy chapter titled Participating in Operational Energy Savings. The structural engineer should distinguish building operating energy for occupant comfort from energy use by a product or assembly. The latter should be included in the product LCA per ISO 21930.
- Material replacement schedules (usually not needed for structural materials). If materials/systems have a design life that is shorter than the design life of the building, the impacts of repair and/or replacement should be included in the LCA.
- Material replacement due to damage caused by natural hazards. See the strategy chapter titled "Performance-Based Design for Material Damage Reduction" for ways to measure repair material savings in buildings resilient to natural hazards.

REFERENCE STUDY PERIOD

The reference study period should be the same for the reference and proposed buildings. If the reference building has a different service life than the proposed

Table 1: Life Cycle Stages and Corresponding LCA Strategies

	PRODUCT stage			CONSTRUCTION PROCESS stage		USE stage (related to building materials)[1]					USE stage (related to building operation)[1]		END OF LIFE		BEYOND BUILDING LIFE[2]	
Module	A1	A2	A3	A4	A5	B1	B2	B3	B4	B5	B6	B7	C1	C2	C3	D
Life Cycle Stage[3]	Raw Material Supply	Transport	Manufacturing	Transport	Construction	Use	Maintenance	Repair	Replacement	Refurbishment	Operational Energy Use	Operational Water Use	Demolish	Transport	Waste Processing	Reuse/ Recycling/ Recovery Potential
Structural Material Quantity Reduction																
Structure as Finish																
Other Nonstructural Quantity Reduction																
Performance-based Design for Material Damage Reduction																
Using Alternate Structural Systems																
Impact Reduction of Functionally Equivalent Material																
Sourcing Salvaged Materials																
Design for Deconstruction																
Contributing to Operational Energy Savings																

1. Portion in parentheses is not part of EN/BS 15804 and is provided here for clarity only.
2. Beyond Building Life is also known as Beyond System Boundary.
3. Shading indicates the life cycle phases during which differences in environmental impacts would be expected for a given strategy.

building, the use-phase impacts of the reference building may be prorated accordingly. The study period is different from the building design service life. Refer to the guiding WBLCA standard, rating system, or code for the required reference study period.

Notes

ASTM E2921 requires a minimum design life of 75 years. LEED v4 requires a minimum design life of 60 years. ASHRAE 189.1 presents a table that shows the design life by building type, although for most buildings it is 75 years. Table 2 of the Athena WBLCA Guide provides further comparison of the reference study period requirements among the green building programs. When project specific design life criteria are not known for the reference and proposed buildings, this WBLCA guide recommends using the 75-year minimum design life for comparative study purposes.

IMPACT INDICATORS

To achieve a valid comparison, the impact indicators used in the analysis should be consistent between the reference and proposed buildings. The impact indicators should provide a statistically similar level of accuracy in representing the material and energy flows to and from the structure throughout the scope of life cycle stages included in each study. The user is referred to the relevant green building program for the impact indicators that should be employed. Improvement targets (and allowable exceedance) for impact indicators vary by green building program. Units for impact indicators should remain consistent between the reference and proposed building studies.

Potentially relevant impact indicators for WBLCA are shown as follows, although this list is not exhaustive (TRACI 2.1 user's guide):

- Acidification of land and water [moles H+ / kg SO_2 (short ton SO_2 equivalent)];
- Eutrophication [metric ton nitrogen (N) / short ton N or kg phosphate $(PO_4)^{-3}$];
- Global warming potential [metric ton CO_2 equivalent / (short ton CO_2 equivalent)];
- Depletion of stratospheric ozone layer [kg CFC^{-11} (short ton CFC^{-11})];
- Formation of tropospheric ozone (smog) [kg $NO_{x,}$ kg O_3 / short ton O_3];
- Human health particulate [PM2.5 equivalent];
- Human health cancer, non-cancer, and ecotoxicity;
- Land use, and
- Water use.

EPDs also require the product manufacturer to report supplemental information, which may include these additional environmental indicators (ISO 21930):

- Nonrenewable material resource consumption,
- Renewable material resource consumption,
- Nonrenewable energy consumption,
- Renewable energy consumption,
- Net water use,
- Nonhazardous waste generated, and
- Hazardous waste generated.

FUNCTIONAL EQUIVALENCE

For the proposed building to be considered functionally equivalent to the reference building, it should meet or exceed the minimum performance standards, such as loading, life safety, and serviceability requirements (i.e., deflection, vibration, durability, noise transmission) of the reference building. The proposed building may exhibit improved performance over the reference building as long as the WBLCA includes the life cycle impacts from the materials used to achieve the improved performance. However, the proposed building may not exhibit decreased performance than the baseline. In all cases, if one material, product, or system has a different service life than another, the impacts of maintenance, repair, or replacement should be included in the analysis.

Notes

Allowing a difference between the proposed and reference building designs is a deviation from traditional LCA definitions of "functional equivalence." When using LCA to compare designs, "functional equivalence" would require the two alternates to strictly conform to the same design criteria. The authors of this guide have proposed this departure to encourage project teams to design structural systems that perform above the minimum levels set forth in standard codes. This deviation is in line with the spirit and intent of LCA, which was created to aid in the improvement of product designs and processes. In contrast, a deviation towards decreased performance and falling below code requirements is not in the spirit of LCA, nor allowed in most green building programs.

The person responsible for the WBLCA should verify that the reference building and proposed building structures are functionally equivalent. More specifically, two building designs can be deemed functionally equivalent if all conditions and constraints in the design criteria of the proposed building design are also met by the reference building design. These conditions and constraints should include but are not limited to the following parameters commonly found in the structural design criteria, where applicable:

- Superimposed dead loads,
- Live loads,
- Snow loads,
- Wind loads,
- Soil conditions,
- Seismic performance,
- Durability exposure class,
- Fire ratings,
- Deflection limits,
- Vibration limits,
- Clear heights,
- Bay sizes,
- Service life, and
- Thermal mass.

In addition, parameters from ASTM E2921 and the green building program relevant to the project should also be included within the definition of functional equivalence. For instance, ASTM E2921 requires that parameters such as orientation, size, location, and program be kept the same. Since these parameters are typically part of the architectural design criteria, they are inherited in the structural design criteria.

While some of the parameters listed above are within the scope of the structural engineer to ensure equivalence, several of the parameters—such as acoustic performance or the fire rating of a total assembly that includes nonstructural materials—may require coordination with other design team members. Differences are allowable in dead load and other traits inherent to the systems themselves, such as fire rating, acoustical properties, and thermal performance, as long as the design criteria of the proposed building design is met by the reference building design.

Example

A steel structure can be compared to a concrete structure even though the steel structure self-weight (and thus seismic weight) is lower than the self-weight of the concrete structure if all else not related to the self-weight is the same, because that difference is only due to the choice of system.

When feasible, the values associated with all design criteria should match between the two designs. However, since an exact match is often not possible (such as when slight differences in the structural system leads to minor differences in floor area, bay size, or clear height), the project team may use professional judgment or perform a sensitivity analysis to show that the effect on material quantities and/or environmental performance would be negligible.

Notes

A structural engineering rule-of-thumb that could be applied is that of "±5 percent," which is used in practice to indicate that numerical differences are negligible. This number should be adjusted in proportion to other factors, such as

- Variability of the design criteria parameter,
- Effect on results, and
- Margin of targeted improvement sought.

In all cases, no attribute of the proposed design may be so inferior to the reference building design that it no longer meets the minimum design criteria, and all materials used to achieve functional equivalence between the systems should be included.

Example

If the reference building uses a concrete structure which inherently achieves its required fire rating, then a proposed building using a steel system should include the fireproofing necessary to achieve the minimum fire rating required for the reference building.

Furthermore, where in the spirit of the requirements in ISO 14040 and 14044 the project team should clearly define the functional unit of study, in a comparative LCA the team should likewise document why the two designs are functionally equivalent and identify any limitations of the study. This is often required in the green building program and is also good practice.

TEAM MEMBERS

The WBLCA preparers should involve all members of the project team who make decisions regarding the materials to be included in the assessment. Project team members responsible for material takeoffs for other purposes (e.g., cost estimating or building information modeling) should be engaged to participate in the WBLCA to improve the accuracy of the materials quantities used in the assessment. The WBLCA strategies outlined in Part II may necessitate the involvement of design team members with specific expertise. For example, when the assessment includes the operating energy-use stage, the project team should engage the energy professional on the design team. Typically, green building programs specify that structural and enclosure elements should be included, at a minimum. Therefore, the team members participating in the life cycle assessment should include at a minimum the members responsible for the enclosure and the structure.

STAGES OF DESIGN

The American Institute of Architects "Levels of Development" are commonly used to define the stages of the design process. In this book, "early design" refers to any point from Concept/Programming or Schematic Design to early Design Development, and "final construction documentation" refers to any time after the end of Construction Documentation, including the end of production of the As-Built drawings. Comparative WBLCA using final construction documents or as-built documentation is the most accurate, as these documents most closely represent the structure as it will be constructed. However, WBLCA comparisons during the concept stage or earlier stages of design offer the greatest opportunity to inform design choices such as materials, systems, or specifications. The user should follow the rules of the green building program if it dictates the stage of design at which the comparative WBLCA should be performed. If it does not, then this guide recommends starting the comparative WBLCA based on early designs and then updating the assessment and submitting results for comparative claims after the final design or as-built set of project documents are complete.

Notes

While the green building rating system or standard will most likely require that the final analysis results reflect the state of final design or construction, the project team will often find that starting the analysis early, when the opportunity to influence design and specification decisions is greatest, will better position the project to attain environmental impact reductions. Starting the analysis well before the team needs to document the design for green building programs is already a lesson learned in the practice of energy modeling.

DATA TO GENERATE A COMPLETE BUILDING DESIGN

In creating a reference building, users will inevitably need additional design data to complete the option that was used to start the reference building. These data should be composed of quantities and processes that reflect standard manufacturing and construction practices and may be based on a compilation of industry-standard materials and practices or information derived through an individual's own practice.

Industry standards can be obtained from averages published by third-party sources, such as those listed in the Examples of Acceptable Sources. When regional practices depart from national standards, the user should adjust quantities in keeping with their own experience and evidence of regional practices. The "Structural Material Quantity Reduction" chapter covers data sources for material quantities. Data sources for industry-average material impact factors are covered in the chapter titled "Impact Reduction of Functionally Equivalent Materials."

LCA TOOLS

The Life Cycle Inventory (LCI) data should be the same for the reference and proposed buildings. If LCI data from different sources are used, the WBLCA should report the sources and discuss the compatibility of the sources. See "Appendix B: LCA Tools" for a list of tools recommended for WBLCA.

Note that while different tools may use the same impact indicators, if the LCI data used differs, the results may not be comparable. Thus, users should only use compatible data sets within a WBLCA, and when in doubt on compatibility, only report information within comparable data sets.

Notes

The easiest way to ensure consistent LCI data is to use the same tool with the same LCA data choices to analyze both the reference and proposed buildings. The available LCA tools range in complexity and ability to customize. The level of detail and customization should be consistent between both buildings and should be documented in the WBLCA report. Note that ASTM E2921 and LEEDv4 require the same tool be used for both the reference and proposed buildings.

DEALING WITH MISSING DATA

Where the LCA tools lack specific materials quantities or life cycle impact data, users should employ professional judgment in choosing the best approximation and justify their choice(s) in a clear narrative. Users can also perform a sensitivity analysis to show that the effect on materials quantities and/or environmental performance would be negligible, as relevant. Generic data or proxy data are often used if specific data are not available.

WBLCA is an emerging practice, and datasets are continuously being developed, refined, and expanded. The lack of specific product information should not discount the ability to perform a WBLCA, although special attention should be given when information is not available.

In such instances, the user should explore the sensitivity of missing data, or data that appears too generic, on the results of the overall WBLCA. For example, if the project team knows foundation concrete volumes may be determined in the field, the team should explore the impact of that uncertainty on the overall analysis. Similarly, if an assembly does not have national average data aggregated by a third party but instead has data from a number of producers, the user could test the sensitivity of the upper- and lower-bound product data on the full WBLCA results. This process is important because some materials, while small in mass, may make significant contributions to the overall WBLCA results. However, if the user can determine that an assembly is not likely to be a significant contributor to the overall result, the best available data can be used as a place holder.

While specific guidance on what constitutes significant sensitivity has yet to be developed, at this time the authors agree that a reasonable initial approach to assess sensitivity is to double the item of interest. If this approach results in a less than 1% change in the overall WBLCA results, it is likely that the overall WBLCA is not highly sensitive to that assembly.

Also, while uncertainty and missing data are inherently part of WBLCA, assemblies that contain significant uncertainty in the quantity or impact measures should not be a central component of the impact reduction strategy.

ADVANCED LCA

At some point during the performance of a comparative WBLCA, the issues may become too advanced or complex for a structural engineer to manage without the aid of an LCA expert. Some examples of situations that may call for an LCA expert include the need to

- Customize data within a building-oriented LCA tool (such as Tally or Athena Impact Estimator) because the material sourcing location(s) for the project are not within the material sourcing region(s) assumed in the tool.
- Create a set of impacts for a material that is altogether not represented in the building-oriented LCA tool used by the project. The user should not combine the impacts of a particular material from one tool with impacts from a different tool that is missing the material without first consulting an LCA professional, starting with the tool provider's technical staff.
- Manipulate data within specific life cycle stages, as required in some of the strategies described in Part II, such as those discussed in the chapters titled "Design for Deconstruction" and "Performance-Based Design for Material Damage Reduction."
- Incorporate data from EPDs into the WBLCA tool.

These techniques will often involve advanced tools such as GaBi and SimaPro, which allow customization. (See Tools in "Appendix B: LCA Tools.")

Reference Building Options

The reference design should be a design that exhibits industry-standard materials and practices for a building that meets the functional requirements of the project. The proposed design should demonstrate a design that intentionally deviates from those industry-standard materials and practices and can be shown as the actual design intention through the latest project design or construction documents.

Notes

The proposed design should generally include feature(s) that deviate from an industry-standard (e.g., the use of lower-impact materials or dematerialization, or a system not commonly used in that application).

In general, comparisons should be made to a reference building that best represents the most appropriate set of specifications for the geographical conditions and functional requirements of the project. While regional data may be available to enable comparisons of regional practices to national or global averages, this guide discourages project teams from setting the baseline to national averages and using regional practices alone as the basis of improvement, where those regional practices would occur by default. Those practices deserve recognition but should not be used for a comparative WBLCA to claim reward in green building programs.

The reference design can be derived in a variety of ways. Regardless of the selected option, care should be taken to ensure that the reference and proposed designs are functionally equivalent. Three basic *starting points* for creating the reference building design are as follows:

- Existing building,
- Building archetype,
- Early design.

Notes

These three options are strongly noted as "starting points" because it is unlikely that any of them is capable of providing a complete and functionally equivalent design that is appropriate to use for comparison at final design. Furthermore, since most green building programs require the WBLCA demonstrate a rather narrow percentage of improvement from the reference design to the proposed design, each design should be sufficiently accurate such that the uncertainty inherent in each design is less than the percentage improvement required.

That said, performing an environmental WBLCA at an earlier design phase offers notable value. For example, comparative WBLCA at early design can ensure the project team discovers opportunities before impactful design decisions should be made. The level of completeness should therefore match the precision required for the purpose of the WBLCA. For instance, for an early rough comparison in which the team is seeking strategies with the most significant impacts, a level of development that falls short of a final design may be acceptable. However, if the user should show a mere 5 or 10% improvement from the reference to the proposed design, both designs should either be modeled with great accuracy to reduce the margin of error and thus demonstrate that the percentage improvement claimed by the team actually derives from the difference in the designs), or the designs should be reported as being "within the statistical margins of error such that differences between the options could not be validated." This can be tested through a sensitivity analysis on the portions of the building the user believes to contain the most variability. (See the section on Dealing with Missing Data for more on sensitivity analysis.)

These options are adapted from the Athena WBLCA Guide. However, where the Athena WBLCA Guide has four options, this book has combined the "building archetype" and "alternative design" into one option for greater simplicity.

OPTION DESCRIPTIONS

Existing Building(s)

A design from an existing building or buildings. This option is most suitable when an existing building (or set of existing buildings) is nearly identical to the proposed building. The project team may use information from the existing building design (and construction, if available) where it matches the proposed building in structural materials, structural system, and design criteria.

Example

An example of using the existing building option as the reference building would be if a client desires the team to replicate a recently designed building located near the building to be designed, such as when a pair or set of functionally, building code-compliant, and stylistically equivalent buildings are situated together on a campus.

Building Archetype

A designed-but-unconstructed building that represents a "typical" building within a defined set of parameters. The project team may use information from third parties such as academic studies, industry publications, contractors, or materials suppliers to construct the Building Archetype. Information from a design firm or owner's portfolio may be acceptable if it consists of a substantial number of comparable projects. The Building Archetype should match the proposed building in structural materials, structural system, and design criteria.

Example

Apply information from a study in which the proposed building is nearly identical to a building in the study. Examples of archetypes can be found in academia and industry, such as

- AISC prototype buildings,
- MIT Buildings LCA Report: *Methods, Impacts, and Opportunities in the Concrete Building Life Cycle,*
- SEAOC *Seismic Design Manual*, and
- Buildings designed based on rules-of-thumb from one's own engineering firm or practice. The Building Archetype using this approach is most robust when the same engineering firm generates both the reference and proposed building designs.

Notes

It is acknowledged that as WBLCA is more widely adopted, engineering firms will develop more robust sources of data internally that can be used for the Building Archetype Option. Once enough data is gathered internally to develop a more accurate reference building design than that provided by academic and industry sources, priority should be given to using internal data sources. In the meantime, when academic or industry sources are used, the practitioner is responsible for validating the applicability of the data for use in defining the reference building.

Early Design

An earlier design for the building under study. This option allows for greater deviation from industry standards or existing precedents but requires evidence that the project team gave serious consideration to the reference design at an earlier design stage of the same project. To use the early design for comparison against the proposed design, the early design should be functionally equivalent to the proposed design (i.e., it should meet the design criteria of the proposed design). Therefore, if the early design criteria have changed slightly and no longer meet the design requirements of the proposed design, the project team should adjust the early design to meet the design requirements of the proposed design.

Example

An example of using the Early Design Option as the reference building would be if two options were compared during the concept phase or later, and the two designs had progressed sufficiently for comparison of their environmental impacts.

COMBINING OPTIONS, FILLING GAPS, AND MAKING FUNCTIONALLY EQUIVALENT

A project team may use a combination of the previously described options to arrive at a reference building design. For instance, the project team might employ all three types as starting points for different portions of a building. For one portion, the team could take an existing building design that is identical in such parameters as area, floor heights, or superimposed loads. For other portions where the bay sizes, loading, and other design criteria are typical, the team might use an available archetype. Then, when the team explored various structural systems for a portion of the building and in the end chose an unconventional system, the team could use an earlier, more conventional design. The last step in this process would be to fill in gaps; for example, in the absence of an existing building, archetype, or earlier designs for components such as the lateral system or foundation, the team might need to use industry benchmark data or even prepare a new design.

Regardless of the option used, care should be taken to ensure that the reference and proposed designs are functionally equivalent, and gaps are substantiated by the references listed under the corresponding option above. In the example given, the team would likely need to further adjust for slightly different design requirements by using professional judgment to scale up or down or to otherwise engineer a rational design. Professionals should perform such adjustments in accordance to best practices already established within their engineering firms.

The three basic options for creating the reference design are shown in Figure 2. These are partially adapted from the Athena WBLCA guide. See the full document at http://calculatelca.com/wp-content/uploads/2014/03/Athena_Guide_to_Whole-Building_LCA_in_Green_Building_Programs_March-2014.pdf for further detail.

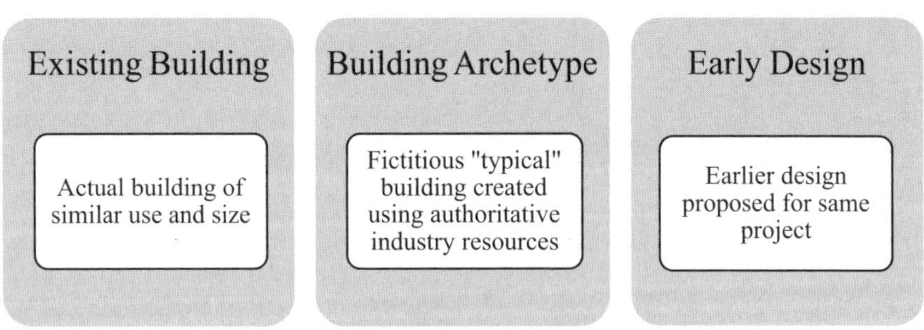

Figure 2: Three starting options for the Reference Building

DOCUMENTATION

Documentation for the comparative LCA should generally include plan and/or elevation drawings for the structural framing, structural details, and/or specifications where required to show a particular impact reduction strategy and a description of how the alternative design differs from the actual design, with additional justifications as appropriate. For each of the options, best practice guidance on documentation is as follows:

- *Existing Building:* Documentation should include the as-built plans and design criteria in addition to explanations and calculations for any modifications necessary to ensure functional equivalence.
- *Building Archetype:* Documentation should include copies or citations (including links if available) to the benchmark data used for comparison. The design criteria for the building archetype should be clearly presented, and any modifications (including calculations or analysis) that were necessary to ensure functional equivalence should be explained.
- *Early Design:* Documentation should include any drawings, specifications, and/or cost estimates that show the similarities and differences between the two systems for the same project. If program, size, orientation, or elements of the design criteria have changed that would impact functional equivalence, the project team should adjust the earlier design and show the calculations necessary to ensure functional equivalence.

Buildings Benchmarking Database

While a robust database of buildings that could be used to generate benchmarks does not yet exist, the SEI Sustainability Committee is encouraged by current efforts underway to create such a database. Once the database is available, project teams will be able to access material information from designs for constructed or constructable buildings that closely match the criteria of the proposed building.

When using such a project database, users could filter the dataset to include only relevant data points. After reducing the dataset, users could take the average of the representative data points as their benchmark for comparison against the proposed design. A project team that employs this method for comparative claims should justify that the data point is representative of the proposed design and then defend any differences.

The advantage of using a benchmark produced from a significant number of data points is that deviations from the proposed building and the spread of data are buffered by the statistical weight of the data.

PhD graduate Catherine De Wolf developed a database with these objectives during her candidacy at MIT from 2014–2017. The "database of embodied Quantity outputs (deQo)" (De Wolf et al. 2016) is now the central tool to collect data for the SE2050 Commitment, a program initiated by the Carbon Leadership Forum. It had input from Architecture 2030 and is currently in pilot by two structural engineering firms. The database can be located at https://deqo.mit.edu. Once the collaborative has launched SE 2050 to the structural engineering community, the SEI Sustainability Committee strongly encourages structural engineers to contribute to the database. More information on these activities can currently be found on the website of the Carbon Leadership Forum.

OPTIONS MATRIX

The strategies presented in Part II of this document indicate the options deemed most appropriate when pursuing a given strategy. Table 2 following summarizes these recommendations across all strategies. As shown in the table, this guide labels these options as "strongly advised," "not advised," or "acceptable, but not the most highly recommended."

This framework may not cover all situations, so project teams should determine which option is most appropriate to each project condition. The project team should use their profession judgment to discern when options that are not advised as follows are more appropriate in special circumstances. If the team employs multiple strategies, several options may need to be combined in the reference building.

Table 2: Reference Building Design Options

Strategy	Existing Building	Building Archetype	Earlier Design
Structural material quantity reduction	**Strongly advised.**	**Strongly advised.**	*Acceptable*, but not the most highly recommended.
Structure as finish	**Strongly advised.**	**Strongly advised.** Ideal. Traditional finishes, but lack of strong data may lead to subjective judgments.	*Acceptable*, but not the most highly recommended. May not have the finishes specified yet.
Nonstructural Material Quantity Reduction	**Strongly advised.** Reusing an existing building may support this strategy depending on specific project scope.	*Not advised.* Unless the archetype contains all systems, this option would be challenging to accurately implement.	**Strongly advised.** Makes a strong case as long as functional equivalence is maintained as the project design progresses.

Strategy	Existing Building	Building Archetype	Earlier Design
Performance-based Design for Material Damage Reduction	*Not advised.* A significant amount of work would likely be necessary to make an existing building match the performance-based design criteria of the proposed design.	*Not advised.* A significant amount of work would likely be necessary to make an archetype match the performance-based design criteria of the proposed design.	**Strongly advised.** This is the only option advised because the user would most likely need to run a damage loss assessment.
Impact Reductions Achieved by Using Alternate Structural Systems	*Acceptable,* but not the most highly recommended. User must establish that existing building is functionally equivalent.	*Acceptable,* but not the most highly recommended. User must establish that existing building is functionally equivalent.	**Strongly advised.** Most applicable to compare functional equivalency.
Impact Reduction of Functionally Equivalent Materials	*Acceptable,* but not the most highly recommended. Conceivable that the user could obtain necessary information from the existing building.	**Strongly advised.** Most suitable to use the same design, except the user would substitute some materials for those better reflecting industry standards (i.e., benchmark data).	*Acceptable,* but not the most highly recommended. Conceivable that the user could show how the earlier design specified different materials.
Incorporating Salvaged Materials	**Strongly advised.**	**Strongly advised.**	**Strongly advised.** Most applicable to compare functional equivalency.
Design for deconstruction (DfD)	*Acceptable*, but not the most highly recommended. Could use an existing building for information on DfD connections, wastage, and likelihood of reuse, but otherwise the reference building should match the as-design, except for the DfD elements.	**Strongly advised.** The archetype would be composed of industry benchmarks for all aspects of both the reference building and as-design. The reference building should have the same system and materials as the as-design.	**Strongly advised.** This is the only case in which one may deviate further in materials or systems from the as-design. Should provide evidence that an earlier design did in fact consider a more dissimilar material or system than the proposed building.

(continued)

Table 2: Reference Building Design Options (Continued)

Strategy	Existing Building	Building Archetype	Earlier Design
Participating in Operational Energy Savings	*Not advised.* This option would require energy professional guidance.	*Not advised.* This option requires energy professional guidance.	**Strongly advised.** This is the only option advised because to account for how structural materials or form affect energy performance (thermal mass, shallower spandrel beams), the user would need to run an energy analysis specific to the building.

REFERENCES

Athena guide to WBLCA in green building programs. http://calculatelca.com/wp-content/uploads/2014/03/Athena_Guide_to_Whole-Building_LCA_in_Green_Building_Programs_March-2014.pdf (Accessed Apr. 26, 2018).

Arup. (2010). *Embodied carbon study: Study of commercial office,* hospital and school buildings. Confidential report. The Concrete Centre, London.

Carbon Leadership Forum. (2017). http://swarmdev2.be.washington.edu/2017/02/09/structural-engineers-2050-commitment-initiative/. (Accessed Apr. 26, 2018)

De Wolf, C., Yang, F., Cox, D., Charlson, A. (2016). "Material quantities and embodied carbon dioxide in structures." *ICE J. Eng. Sustain.* 169(ES4), 150–161, DOI: 10.1680/ensu.15.00033. http://www.icevirtuallibrary.com/doi/abs/10.1680/jensu.15.00033 (Accessed Apr. 26, 2018).

Hsu, S. (2010). *Life cycle assessment of materials and construction in commercial structures: Variability and limitations*, Massachusetts Institute of Technology, Cambridge, MA.

Ochsendorf, J., Norford, L. K., Brown, D., Durschlag, H., Hsu, S. L., Love, A., et al. (2011). *Methods, impacts and opportunities in the concrete building life cycle*, Massachusetts Institute of Technology, Cambridge, MA.

Simonen, K. (2014). Life cycle assessment. https://www.routledge.com/Life-Cycle-Assessment/Simonen/p/book/9780415702423 (Accessed Apr. 26, 2018).

PART II

Strategies

Structural Material Quantity Reduction

SCOPE

When demonstrating material quantity reduction, the reference building design should be equivalent to the proposed building, except where the proposed building utilizes less material (i.e., the structural engineer has chosen a more materially efficient system composed of the same basic materials). Within this strategy for the reference building, reducing the building's structural material quantities is achieved by using a nonconventional design or more optimized system within the buildings' same material.

This chapter discusses reduction strategies for whole structural materials only; for other related strategies, see the following chapters:

- *Impact Reductions Achieved by Using Alternate Structural System* describes exceptions such as when an earlier design option for the same project that meets the same programmatic requirements used a dissimilar material or when the proposed building utilized a different system/material than the industry standard for the building type.
- *Impact Reduction of Functionally Equivalent Materials* is most applicable for changes in the constituents of a material (i.e., cement in concrete).
- *Performance-Based Design for Material Damage Reduction* discusses damage reduction, which reduces the need for additional materials through the life of the building (e.g., damage in a catastrophic hazard event).

In all cases, the structural system of the reference building should maintain the same functional equivalence.

BILL OF MATERIALS ADJUSTMENTS

This strategy is verified by comparing the bill of materials (BOM) for the reference building to the proposed building. (See the Special Considerations section following for further explanation and examples categorized by material-specific strategies.) When determining the BOM for the reference building, the following are acceptable practices/aids:

- Industry rules of thumb developed through years of practice can be used to estimate material quantities for structures. This practice is acceptable as long as the project team provides the reviewer with these rules of thumb.
- Design or specification guidelines produced by trade organizations or professional associations may be utilized to determine material quantities.
- Industry surveys published by construction-sector research organizations are acceptable to determine material quantities.
- BOM databases such as the Athena Impact Estimator's assemblies or others may be utilized. Internal company databases may also be used as long as the reviewer can access the database to confirm its validity.
- Any other approach that is justified by the engineer of record and can be explained to the green building program reviewer is acceptable.
- The material quantities calculated for the reference building should be determined from the documentation. When determining the BOM, the proposed building should be at the same level of design (i.e., schematic, design development, construction documents) as the reference building.

  Examples of Acceptable Sources

  Design or specification guidelines that can be used to determine material quantities include (but are not limited to) light-gauge metal-stud design tables, ASCE design guides, and American Institute of Architects design guides.

  Industry surveys published by construction-sector research organizations include *RS Means*, which has proportionate quantities tables as part of the reference tables, and *The Architect's Studio Companion Rules of Thumb for Preliminary Design*, which has capacity charts and tables for various structural systems.

  Database of Embodied Quantity Carbon (DeQo); see further information in the "Reference Building Options" chapter.

LIMITATIONS

When determining the BOM for the reference building, the project team should ensure that the building's bay size, programming, and loading is as equivalent as possible to the proposed building or industry standards. The user should not indicate a reduction in material quantities for the proposed systems when the industry standard is a more efficient system (e.g., if post-tensioned slabs are the standard for multistory residential construction, the team cannot utilize a mildly reinforced concrete slab for the reference building and a post-tensioned slab for the proposed). This strategy may be pursued in combination with any of the other strategies to minimize the life cycle impact of the proposed building over a reference building.

SPECIAL CONSIDERATIONS

Material-Specific Strategies

Options for achieving material quantity reductions for the proposed building have been grouped by building material below. The intent is to demonstrate material quantity reduction in the proposed building systems that are of the same material as the reference building systems. If the alternative materials or systems used in the proposed building have a shorter lifespan than the design lifespan of the proposed building, the WBLCA should include impacts for the repair and/or replacement of the material or system. The strategies listed following are not guaranteed to lead to reduced life-cycle impacts and need to be verified by the WBLCA.

Concrete

Options for demonstrating material quantity reduction in concrete systems include:

- Utilizing post-tensioning, except when post-tensioning is the typical industry practice for the building type. For example, post-tensioned slabs typically require less concrete depth and reinforcing steel than a concrete slab—of the same span, design loading, and concrete mix—reinforced with mild steel.
- Utilizing a voided slabs system (waffle slab, hollowcore, or proprietary voided slab system) can reduce the volume of concrete in a slab compared to a solid slab of the same span, design loading, and concrete mix. The WBLCA should consider impacts from these systems related to the additional formwork and/or voided material.
- Using higher strength concrete mixes than industry standard for the elements under design to reduce the quantity of concrete used to meet same member strength. The WBLCA should account for any increased environmental impacts resulting from the manufacture of the higher strength mix.
- Using higher strength reinforcing steel or using couplers or other bar terminators to reduce the amount of reinforcing steel needed. The WBLCA should account for any increased environmental impacts resulting from the manufacture of the higher strength reinforcing steel as well as for the impacts of the bar couplers or terminators.

Wood

Options for demonstrating material quantity reductions in wood systems include:

- Advanced framing, or optimum value engineering, which utilizes techniques that reduce the amount of wood framing required compared to a conventionally framed wood structure. The main principle in optimum value engineering (OVE) calls for increasing wall stud spacing and aligning wall studs with joists and rafters. This approach also allows a single top plate to be substituted for traditional double top plates.
- Utilizing a hybrid wood system, which reduces the amount of wood in the building but requires additional materials that need to be accounted for. An example of this option is using a steel lateral force resisting system (LFRS) in lieu of a wood LFRS. This approach reduces the amount of wood in the structure, but the added nonwood material should be accounted for.

Steel

Options for demonstrating material quantity reductions in steel systems include:

- Utilizing composite design, increased camber, and castellated or cellular beams to minimize the quantity of steel used. The WBLCA should consider the impacts resulting from the castellation process or any material transportation premiums. In all cases, the WBLCA should provide a narrative to explain the deviation from standard practice.
- Utilizing systems that are more materially efficient than industry standard. For example, use SidePlate, which is a proprietary moment connection design that reduces the amount of steel required for moment frames.
- Utilizing braced frames or damped moment frames instead of standard moment frames as the LFRS to reduce material quantities, where moment frame is industry standard.
- Utilizing long-span deck systems that eliminate intermediate framing.
- Using higher strength steel to reduce the quantity of steel used to meet same member strength. The WBLCA should account for any increased environmental impacts resulting from the manufacture of the higher strength material.
- In utilizing lightweight concrete in lieu of normal weight concrete, the project team can reduce the overall weight of the structure and potentially the amount of material required. However, the WBLCA should consider the more energy-intensive process of producing lightweight aggregate.

Strategies Applicable to Any Material

- Specifying and designing with higher strength materials that reduce the quantity of material used. The WBLCA should account for any increased environmental impact resulting from the manufacture of the higher strength material. (Note: Proxies are acceptable until real data are available.)
- Performing a more rigorous analysis than typical industry practice (e.g., nonlinear and time-history analysis, use of site-specific studies to reduce design loads, eliminating over-conservatism) to reduce the quantity of material

used to meet code requirements. This more advanced analysis could lead to changing the structure's boundary conditions (e.g., providing column fixity at the building's lateral force-resisting system to reduce the amount of steel required), but in doing so, the user should justify why such an analysis was not performed on the reference building.

- Using performance-based design to meet owner's performance objectives, which reduces materials in comparison to code-based design.

Structure as Finish

SCOPE

This chapter addresses the WBLCA requirements for demonstrating reduced environmental impacts by reducing nonstructural materials, which can be achieved through deliberate structural design choices and/or exposing structural elements, as opposed to the more traditional practice of covering structural elements with finish materials. For other related strategies, see the following chapters:

- *Structural Material Quantity Reduction* discusses reducing structural materials to reduce impacts.
- *Nonstructural Material Quantity Reduction* discusses reducing nonstructural materials but not due to using structure as finish.
- *Impact Reduction of Functionally Equivalent Materials* discusses optimization of constituent materials.

Furthermore, this chapter does not address how to implement strategies for exposing structure since exposing structural elements can be complex and can be based on subjective reasoning (appearance) and/or objective reasoning (cost, fire resistance, weather resistance, etc.), which requires an integrated design approach that is beyond the scope of this book.

BILL OF MATERIALS ADJUSTMENTS

When determining the BOM for the reference building, project team members may use the references recommended in "Structural Material Quantity Reduction" where they also offer guidance for nonstructural materials. When using this strategy, the team should include nonstructural materials in the BOM for both the reference and proposed buildings. This strategy is likely to reduce the BOM within the nonstructural portion of the BOM.

LIMITATIONS

When the reference building design demonstrates environmental impact reduction by exposing structure (reducing or eliminating finish materials), the proposed building and reference building designs should be functionally and building code compliant, except that the proposed building unconventionally exposes portions of the structural elements. The reference building should include non-structural materials when those finishes are typically applied.

Example

In a warehouse building, a reference building with carpet should not be compared to a proposed building without carpet, because a warehouse building typically does not contain carpet. The only allowable exception is if the project team is able to show that an earlier design of the project included the finish material (e.g., carpet) in official project documentation, such as in a cost estimate or in an issue for permit review.

For the structural systems to be considered functionally equivalent, they should not only meet loading requirements but also serviceability and life safety requirements, including fire, deflection, vibration, durability, maintenance, and noise transmission, as well as any other building code, functional, and programmatic requirement of the building. The WBLCA should discuss any changes in the environmental impacts of structural elements as a result of modifying the exposed structural elements to meet the code and functional requirements for the building.

SPECIAL CONSIDERATIONS

To qualify as an acceptable proposed building with exposed structural elements, the building should be functionally equivalent to the reference building with structural elements covered with finishes, where the application of those finishes is industry standard. In addition, the WBLCA should account for differences in service life and maintenance regimes to render the buildings functionally equivalent. In many cases, an exposed structure requires additional maintenance compared to a covered structure, although in some cases the opposite is true. Regardless, the WBLCA model should consider the impacts of maintenance or replacement through the reference study period of the LCA.

General Issues that Apply to All Structural Materials

One way to reduce the environmental impacts of a building is to reduce the amount of finish materials by exposing the structural elements. Many building designs incorporate exposed structural elements, both on the interior and exterior of a building. The decision to expose structural elements is complex and can be based on subjective reasoning (appearance) or objective reasoning (cost, fire resistance, weather resistance, etc.). Reducing environmental impacts by eliminating finish materials can be an additional rationale.

Many of these decisions are not driven by structural engineers; however, the project team should be aware of the decision to expose a structural element since it may have an effect on structural material properties and design details.

Many structural assemblies rely on special finishes or treatments to meet building code requirements for fire resistance. Structures built as "noncombustible" Type I or II framing or those that exceed the height and square footage limitations for Types III through V per the International Building Code typically require additional treatment for steel and wood, such as sprayed-on fireproofing, fire-resistant paint, fire-resistant gypsum wall board, plaster, or other fire-rated system. Although the finishes are not part of the assembly designed to resist structural loading, without these finishes the assembly would not likely meet code requirements.

Weathering and durability for structural elements exposed on the exterior is also a consideration. Some structural materials and products can resist weathering without additional finishes, but most require some special treatment that should be taken into account when conducting the WBLCA.

Some structural assemblies require additional finishes to meet building code or functional requirements of the project to control deflection and vibration. For example, floor systems often rely on interior partitions for deflection and vibration control. In other cases, steel and wood floor and wall systems rely on additional finishes to meet noise transmission ratings. Concrete and masonry typically do not require additional finishes, but in some cases may need additional finishes to meet these requirements. Regardless, for a proposed building to be functionally equivalent to a reference building, these considerations should be taken into account.

Material-Specific Considerations

For all materials, the project team should consider how finishes contribute to building serviceability when attempting to eliminate them. For serviceability issues such as deflection, vibration, or noise transmission, most floor systems rely on finishes to meet serviceability criteria.

Concrete

Generally, concrete structural elements can be functionally equivalent with or without finishes. Floors, walls, columns, and other elements may more often inherently meet building code and project requirements for fire, durability, deflection, vibration, and noise transmission. However, structural concrete that is intended to have a finish material may not be suitable for exposed applications in terms of appearance. Special techniques for improving appearance, such as formwork treatment, coloring, texture, polishing, or formulation (self-consolidating concrete), have the potential to increase environmental impacts. However, for the most part, these treatments would likely have relatively low impacts compared to traditional finish materials. Regardless, any treatment of concrete elements to meet the life safety, functional, and programmatic requirements of the building should be taken into account when performing the WBLCA.

Masonry

Generally, masonry meets building code requirements without additional treatment or finishes. For example, a masonry wall may meet the requirements for fire, durability, deflection, vibration, and noise transmission without additional treatment of coverings. Thus, the project team could design the reference building with finishes and the proposed building without finishes or with minimal finishes such as paint. However, in many cases, masonry walls are traditionally left unfinished, and in these cases, it would be difficult to justify designing a reference building with finishes over the masonry walls. Some decorative masonry could have a higher impact than nondecorative masonry, which would have to be taken into account when conducting the WBLCA.

Steel

Creating a building design with exposed structural steel commonly reduces finish material quantities. The team should examine the model building code to determine the size and height of structures that can be built without fire protection. Steel that will remain exposed to the elements in its final condition should be treated (e.g., via high-performance paint system or hot-dip galvanizing) to achieve the performance and service life required.

One example is to include exterior steel framing elements that would be covered with a drop-ceiling tile in the reference building and exposed in the proposed building.

Another example would be structural steel assemblies exposed on the exterior of a building. Some structural materials require finish coatings or coverings to resist weathering (corrosion, rotting, etc.). Steel floor members on the exterior of a building are often covered with a finish material on the top (floor) and bottom (ceiling) or painted, galvanized, or treated to resist weathering. The project team will need to include any necessary materials such as these to meet functional equivalence requirements between the reference and proposed buildings.

Wood

Exposed wood structural elements may require additional treatment to meet the life safety, functional, and programmatic requirements defined for the building. Because there are examples of heavy timber construction with exposed structural elements, the project team could design a reference building with uncovered timber and an exposed structure, provided it met fire code limitations. However, in many cases, timber structural elements are traditionally left unfinished, and therefore it would be difficult to justify designing a reference building with finishes over the timber elements. Exterior exposed wood elements would require weather treatment, and the impact of the treatment would need to be taken into account. Conventional wood framing or engineered wood products, however, would rarely be exposed. Fire code requirements, durability requirements, deflection limitations, and vibration and noise transmission considerations mostly rely on additional finishes to meet code and project functional requirements.

Treatment of Exposed Structure in Green Building Programs

The various green building rating systems and standards treat finish materials in different ways. For a summary of these requirements, refer to Table 3 in the Athena WBLCA Guide (2014). Scope requirements in the three most commonly used green building programs are summarized below:

Green Building Program Scopes

Green building programs that include finishes in the scope of the WBLCA allow the use of this strategy. In contrast, any green building programs that explicitly exclude finishes from the scope of the WBLCA would prohibit use of this strategy since the finishes cannot be included in the reference building.

LEED v4

LEED v4 contains requirements for defining the scope of WBLCA. While the language regarding inclusions and exclusions of certain finishes is somewhat ambiguous, the Athena WBLCA Guide states that finishes on floors, ceilings, and interiors walls are optional. If optional elements are included in the LCA, the functional equivalence must be maintained between the reference building and the proposed building.

Green Globes

For Green Globes certification, credit is given when the WBLCA demonstrates that a proposed building has lower impacts than a reference building. Green Globes allows inclusion of floor and ceiling finishes, and thus it is possible for the reference building to have structural floor assemblies with finishes and for the proposed building to have structural assemblies without finishes, provided they are functionally equivalent.

International Green Construction Code

The IgCC imposes a jurisdictional requirement for WBLCA when selected by the adopting jurisdiction governing the project. The IgCC places clear limitations on what should be included in the LCA, and specifies for the inclusion of "interior walls, floors and ceilings, including interior . . . finishes." Therefore, one could analyze the reference building with architectural finishes and analyze the proposed building using exposed structure.

Nonstructural Material Quantity Reduction

SCOPE

This chapter addresses how project teams can use WBLCA to demonstrate reduced environmental impacts resulting from integrated design not related to structure as finish. It also provides guidance on how to calculate reductions in environmental impacts that result from integrating design of the structure with building envelope, building finishes, and/or building mechanical/electrical/plumbing (MEP) systems. For other related strategies, see the following chapters: "Structure as Finish" discusses nonstructural material reduction due to exposing structure, and "Participating in Operational Energy Savings" discusses modeling requirements when including operational energy reductions.

BILL OF MATERIALS ADJUSTMENTS

Minimization/Elimination of Nonstructural Materials

In demonstrating reduced environmental impacts, often as a result of integrated design, the WBLCA should show that the reference building design is equivalent to the proposed building, except for the reduced material in the envelope, finishes, and/or MEP systems when the structure provides certain functions for these systems instead.

Example

The use of passive solar design techniques is one example whereby the structure is used to store and exploit the sun's energy to reduce the overall energy demand of the building and the need for MEP systems (e.g., heating and lighting). Another example is to rework column grids and floor spans to reduce floor-to-floor heights, thereby reducing the overall height of the proposed building and further reducing the quantity of building envelope, interior wall finishes, and MEP systems (ductwork, pipe, and electrical conduit). In this example, the overall building volume would also be decreased, potentially also reducing energy demand.

SPECIAL CONSIDERATIONS

The WBLCA should account for the reduced impacts of integrating the building's systems over the life of the building. To show impact reduction for the proposed building over the reference building, the WBLCA needs to consider synergies among the building's embodied and operational impacts. The reference building design should consider industry-standard impacts for the building type as they pertain to the envelope, finishes, and MEP systems, if within scope of the green building program. If this strategy is the only strategy pursued, the reference building should be of the same material as the proposed building; however, it may be pursued in conjunction with other strategies.

Integrated Design Strategies

- The reduced impacts of integrating the building's systems should be accounted for over the life of the building (e.g., a higher performing building envelope may have a high embodied energy, but the building's savings on heating and cooling demands over the life of the building may offset the envelope's initial impact).
- Thermal mass strategies that reduce a building's heating and cooling demands may be utilized for reducing the impacts of the proposed building, but an energy model should be employed. Similarly, increasing floor heights may allow daylight penetration and reduce electric lighting installed. These effects should be accounted for by including the use phase in the WBLCA. (See the chapter on "Participating in Operational Energy Savings.")

Envelope and Other Nonstructural Structural Elements

- Reducing the structural floor system depths to reduce the façade, interior finish, and mechanical system height/area/length.

Performance-Based Design for Material Damage Reduction

SCOPE

This chapter provides guidelines for quantifying building material damage from natural hazards. Building damage from predictable, reoccurring hazards can be reduced over the life of the proposed building through improved performance under environmental loads beyond those prescribed by code.

Performance-based design involves evaluating how a building is likely to perform given the potential hazards it is expected to experience. This method takes into consideration the uncertainties inherent in both quantifying potential hazards and in assessing the actual building response. This design approach is an alternative to the analysis methods and construction practices prescribed by building codes that typically focus on limiting the loss of life, with less consideration to other consequences. Performance-based design can provide better building resiliency when performance objectives are aimed at reducing building damage, because verification of performance occurs through a more rigorous and robust analytical and probabilistic method rather than a prescriptive, code-based approach.

Notes

One common resiliency strategy is to improve building performance during seismic events, thereby reducing delays to building reoccupation, reducing repair costs, and reducing short-term environmental and social impacts on the recovering disaster area.

BILL OF MATERIALS ADJUSTMENTS

The WBLCA should compare the replacement and repair materials needed for the use stage (B3 Repair, B4 Replacement) of the baseline building against materials saved by a more resilient proposed building. The WBLCA preparer should use a reliable source to estimate related damages. The assessment should quantify the difference in material quantities needed to make the proposed building more resilient than the baseline building.

Performance-based design may require additional materials to create a more robust structure in the construction phase but aims to reduce the probability of building material damage, thereby reducing the need for repair labor and materials during the maintenance phase of the building's life. Other performance-based design strategies may reduce total materials but may then have higher initial costs. The strategies described following are examples that may involve higher initial costs.

For seismic performance-based design:

- Increased sizes of lateral-load-resisting members,
- Secondary lateral-load-resisting systems such as outriggers or frames,
- Base isolation systems,
- Supplemental dampening systems,
- Alternative foundations,
- Ground improvements to mitigate settlement or lateral spreading,
- Increased jointing of cladding systems, and
- Added material for improved resilience of finishes.

For flood-resistant design:

- Raise the building grade or floor elevation above the floodplain,
- Move impactful building elements such as the HVAC and generators out of the base flood elevation, if the rating system or standard allows these to be included in the scope of the WBLCA,
- Water-resistant materials, and
- Breakaway walls.

For wind-resistant design:

- Resilient roofs,
- Debris-resistant facades, and
- Select impact-resistant glazing and other exterior finishes.

For wildfire-resistant design:

- Noncombustible roofing and cladding, and
- Exterior sprinklers.

Notes

Of the previous hazards, seismic, flood, and wind hazards have established methodologies for determining recurrence intervals and quantifying damages. Damages from hazards such as forest fires may or may not be quantifiable.

The WBLCA may only include damage to finishes and permanent building components where allowed by the rating system or standard. All components should be consistent between the reference building and the actual design.

Notes

The WBLCA will need to include or exclude portions of the building and its contents according to the green building program the project team is following. For example, most rating systems require the WBLCA to cover the complete building envelope and structural elements, including the material components of footings and foundations, and some may also include additional building components such as interior nonstructural walls, ceilings, floors, and finishes.

Although there may be significant damage to permanent nonstructural components that are needed for continued building operation, many rating systems may exclude such items from the WBLCA. Common exclusions are electrical and mechanical equipment and distribution systems, plumbing fixtures, fire detection and alarm system fixtures, elevators, and conveying systems. See the Boundaries and Service Life sections under the General Considerations chapter in Part I for more discussion.

For all components, the WBLCA should provide functional equivalency for both the reference building and the actual design.

The WBLCA should not include external impacts such as economic losses due to downtime or short-term energy and resources consumed for temporary function (occupancy) during recovery.

LIMITATIONS

In demonstrating resiliency through performance-based design, the WBLCA should show that the reference building has the same design as the proposed building, except for components that have been altered to reduce the probability of damage from natural disasters. This measurement should be made through the Early Design Option of the reference building only, where all calculations are performed for two versions of the proposed building. (See the "Reference Building Options" chapter.)

The WBLCA should quantify the materials and energy needed to create the performance-based design. The quantified losses may be measured in dollars but should be limited to the material and energy used to repair or replace the building elements.

Accounting for Damage

For damages resulting from a hazard event predicted to occur during the building life, the WBLCA should incorporate such damages into the use stage (B3 Repair, B4 Replacement from System Boundary Table 1: Life Cycle Stages and Corresponding LCA Strategies) for as many times as the event is predicted. Damages for common events (e.g., 10-year events) may be combined with less-common events (e.g., 50-year events).

The WBLCA should determine the hazard event recurrence interval through a probabilistic assessment, not through deterministic means. The likelihood of the event and ensuing damage should be consistent with the building lifespan used in the WBLCA.

The WBLCA should employ statistical discounting, by annualizing damages or other means to assess impacts of large-magnitude events that have a low probability of occurring during the service life of the building.

Example

For statistical discounting, the maximum considered earthquake (MCE) defined in US building codes has a mean recurrence interval of 2,500 years (i.e., a 2% likelihood of occurring during a 50-year building service life). Thus, 2% of the damage of an MCE event should be included in the life-cycle impacts of a comparative WBLCA.

The hazard criteria should be site specific to the level of detail required by the building code, at a minimum, for both the reference building and the proposed building.

If more accurate data are used for the performance-based design, such as site-specific response spectra or time histories, the WBLCA should analyze the reference building using the same parameters and method as the actual design.

Alternately, structural engineers may employ the Building Archetype Option for the reference building. In this option, the reference building may not be any more susceptible to damage than allowable by code, and a structural engineer should use the same methods described for using the Early Design Option to compare performance.

Accounting for Resiliency

Resiliency benefits should be limited to the building's defined service life, and benefits are not transferrable to future life cycles. The impacts associated with creating resilient elements or assemblies may not be discounted to reflect their remaining service life, other than standard end-of-life credits such as recycling or salvage.

SPECIAL CONSIDERATIONS

Inventory databases may be derived from material and energy process data or from economic data. If the BOM from the hazard event(s) is estimated in terms of financial losses, the dollar values may be converted to environmental impacts using economic-input-output LCA (EIO-LCA) databases. If other stages are measured using process-based LCA methodologies. EIO-LCA can be used exclusively for the WBLCA use stages documenting the hazard event (e.g., B3, B4), as long as EIO-LCA is used for both the proposed and reference buildings. The authors recommend project teams who wish to use this approach consider enlisting the aid of an LCA professional.

Notes

BOM due to seismic damage is commonly estimated using the Advanced Engineering Building Module of the FEMA/NIBS earthquake loss estimation (a.k.a. HAZUS) methodology or the Performance Assessment Calculation Tool of FEMA P-58. These tools quantify risk in terms of direct financial losses. An example of how this has been implemented can be found in a joint study by the University of Washington and Arup that used the material loss estimation from application of the HAZUS methodology on a 42-story concrete building in a comparative WBLCA (Simonen et al. 2015).

Impact Reductions Achieved by Using Alternate Structural Systems

SCOPE

This chapter sets forth procedures and reference methodologies for the WBLCA preparers to follow when demonstrating an impact savings achieved when the project team selects an alternate structural system and/or changes the configuration of the system.

The project team should design the reference building based on general construction practice for the building type and location. The WBLCA should use standardized datasets typical of representative design from national organizations, when available, to define the reference building. Preference should be given to regional benchmarks over national benchmarks. If such a dataset is not used, the project team should justify why such benchmarks are not appropriate for the project.

Notes

The intent of this chapter is to provide guidance when the user desires to demonstrate material impact reduction in systems for a reference building that employs a different structural system than the selected design. Common examples of this strategy include switching from a concrete system to wood where the concrete system is most conventional for that building type (or vice versa) and switching from structural steel shapes to cold-formed steel framing where one is conventional (but the other is not) for that particular building type in the project region.

In the WBLCA, the reference building is restricted to conventional practice, but the proposed building is not. Project teams are encouraged to explore atypical designs to see if they produce lower environmental impacts.

The team may use the benchmark data to develop a full design, although benchmark data alone are typically not sufficient for the reference building.

SPECIAL CONSIDERATIONS

A key requirement when employing this strategy (i.e., selecting an alternate structural system and/or changing the configuration of the system) is that the project team members explain the reference system, clearly justifying why the improved system is not a typical system for the subject building. Examples include:

- Systems that are typically used in one region but are uncommon in the region of the proposed building. In such cases, transportation impacts should be considered with greater detail to ensure increased transportation impacts do not offset material savings.
- "Novel" systems. In such cases, references should be included substantiating the current reach of the system and the reasons it is considered "novel" for this use.
- "Conventional" systems that may have been an initial recommendation or that, while common in the region, are less common for the subject structure type. The WBLCA narrative should clearly explain the rationale for the switch from the initial recommendation or system.
- Switching systems from an initial/early concept system. If the strategy involves switching systems, both of which are common within the region for the building type, the narrative should demonstrate why the initial system was selected and why the final system was not initially selected. Regional examples of existing buildings that use the initial system should be cited.

If employing this strategy involves sourcing from a specific supplier and location because of improved production methods, the WBLCA should also address any increase in transportation impacts associated with the sourcing scheme.

When changing the structural system involves a change in material, the environmental impact information of the reference building design may come from one of the following:

- Generic materials embedded into commercially available WBLCA software, or
- Industry-wide EPDs compliant with the requirements of the rating system being used and developed following the same product category rule for the life-cycle phases and service life defined for the comparative study.

BILL OF MATERIALS ADJUSTMENTS

Because this strategy involves comparing alternate structural systems, significant differences are expected in the BOM for the reference and proposed buildings.

LIMITATIONS

Because this strategy involves comparing alternate structural systems, there will be significant differences between the reference system and the proposed system. However, both systems should have similar, or improved, BOMs compared to industry benchmarks, or other archetype data. This strategy may be pursued in combination with any of the other strategies.

Impact Reduction of Functionally Equivalent Materials

SCOPE

This chapter sets forth procedures and reference methodologies for WBLCA preparers to follow when demonstrating an impact savings by making changes in the sourcing or manufacturing of a specific material rather than changing the system and using different materials or changing the configuration of materials.

Notes

The intent of this chapter is to provide guidance when the user desires to demonstrate material impact reduction in systems for a reference building that are the same material as the systems in the proposed building. Common examples of this strategy are optimizing concrete mix proportions to produce a functionally equivalent concrete, but with less impact, or sourcing another material or commodity product in such a way that it provides functional equivalence with less impact.

SPECIAL CONSIDERATIONS

The project team should design the reference building using general construction practices for the building type and location. When available, standardized benchmarks from national organizations should be used to define the reference, and preference should be given to regional benchmarks over national benchmarks.

The environmental impact information on compared alternatives should come from one of the following:

- Options embedded into commercially available WBLCA software such as building products from constituent materials available in the WBLCA software, or

- Product EPDs compliant with the requirements of the rating system being used and developed following the same product category rule for the life-cycle phases and service life defined for the comparative study.

If employing this strategy involves sourcing from a specific supplier and specific location due to improved production methods, the WBLCA should also address any increase in transportation impacts associated with the sourcing scheme.

If using manufacturer-specific EPDs, the WBLCA preparer should ensure that the EPD used for the proposed "improved" material follows the same scope and assumptions of the material data built into the WBLCA tool. In such cases, it is strongly recommended that the user work with the material supplier and LCA tool provider to include the product in the LCA tool instead of the user adding assembly impact information outside of the LCA tool.

BILL OF MATERIALS ADJUSTMENTS

The BOM may not be affected when the team implements this strategy in isolation. However, this strategy may frequently intersect with the material quantity reduction strategy.

MATERIAL-SPECIFIC STRATEGIES

Strategies to reduce material impacts have been grouped by building material.

Wood

When demonstrating impact reductions in wood elements, the user should

- Place particular consideration on the inclusion of transportation impacts. Depending on geography, certain species or products may not be locally available and transportation can be a large contributor to the total impacts of wood members.
- Ensure that any process impacts required to achieve the engineering properties of the final material are accounted for (e.g., kiln drying).
- Use specific species data. Some commercially available software packages group wood entries as an average of the impacts of multiple species (i.e., Douglas fir and Southern yellow pine). However, different species generally grow in different regions and have different impacts. If a strategy is to demonstrate impact reduction though specific species data, data for the selected members should be compared to regionally appropriate species data instead of national mix data.
- Remain aware of the following topics that are beyond the scope of this document:

  - Using sources that are already compromised and may cause larger risks if not harvested (beetle kill), and
  - Differences in the impacts of species depending on specific methods of growth and harvest (e.g., impact of fertilizers on overall embodied impacts and regional forestry/foresting practices).

Steel

When demonstrating impact reductions in steel elements, consider that

- Current industry-average EPDs for fabricated structural steel shapes, hollow structural sections, and plates are available from AISC.
- Mill-specific EPDs, or LCA data in the LCA tool used, should follow same product category rules and boundary conditions as the LCA data used for the reference building. In order to prove an advantage of a product or process specific to an individual mill, the project team should use EPDs from specific manufacturers. These EPDs are currently under development.
- For the case of using a higher strength steel, such as higher grades of rolled shapes or rebar, the team should seek to reference EPDs (or LCA data in the LCA tool used) for both grades, following the same product category rule, to determine differences in impacts per functional unit.

Notes

At this time, industry-average impacts are not available for grades of steel other than the preferred material specifications for hollow structural sections and structural shapes. Individual manufacturers may produce EPDs for varying grades of the same material in the near future. Currently, the impacts accrued from producing different grades of steel are not expected to differ considerably when produced by the same mill. Overall impact reduction will derive predominantly from using less material of a higher strength steel, not from its production process. Impact differences due to specific mill and process differences would be captured in mill specific EPDs.

The WBLCA should address the accompanying material reduction in metals or other materials as described in the "Structural Material Quantity Reduction" chapter.

Concrete

When demonstrating impact reductions in concrete mixtures, note the following:

- The WBLCA should use benchmark mix proportions from *NRMCA Member National and Regional Life Cycle Assessment Benchmark (Industry Average) Report* for the reference building and use regional numbers unless the project team provides a compelling reason as to why the national averages are more applicable.

- If reference mixes do not contain supplementary cementitious materials, the team should provide a detailed description as to why such a mix is appropriate as a reference.
- In most cases, reference mixes should use the 28-day benchmark as a reference strength. For cases where the 28-day strength is not the governing specified strength, such as post-tensioning, the WBLCA may either use benchmark mixes for "high early strength" mixtures or, if such references are not available, 28-day benchmark mixes that match the 28-day strength of the mix. For slower setting mixes such as those with a high fly ash content, it is allowable to compare 56-day strength mixes to the 28-day strength mixes of more conventional concretes.

Notes

Permitting 56-day strengths is to encourage consideration of lower strength cement mixes, which may require the team to set a longer than typical time for the concrete to reach compressive design strength (i.e., greater than 28 days).

For mixtures with special requirements, such as enhanced durability requirements where benchmarks are not available, the team may justify the use of other reference mix proportions.

The National Ready Mixed Concrete Association does not include a 10,000-pound-per-square-inch mix, but the WBLCA may use comparisons to past project mixes.

LIMITATIONS

Deviations in Material Quantities

If this strategy is the only strategy employed, the BOM should generally be the same in both the reference and proposed building, with the notable exception related to the use of higher steel grades. However, this strategy may work in coordination with other strategies, such as Structural Material Quantity Reduction, which may result in differences in the overall BOM.

Incorporating Salvaged Materials

SCOPE

When working toward sourcing salvaged items/content, the reference building design should be equivalent to the proposed design but use virgin materials in lieu of any salvaged/reclaimed materials. The structural system of the reference building should maintain the same functional equivalence in all other aspects. It should also be of as similar material as reasonably possible to the proposed design. In taking credit for using salvaged materials, the professional conducting the WBLCA may subtract or reduce the environmental impact burdens in modules A1-A3 (product life cycle stage; see Table 1) for the portion of materials that are salvaged (i.e., cradle-to-factory gate impacts may be neglected when using salvaged material).

Necessary Refabrication

In cases where the salvaged material will be transported and refabricated, the WBLCA should assign new impacts from modules A2 and A3. In general, the only impacts carried by the materials should be from modules A4 and A5 (design and construction process stage).

> **Notes**
>
> In cases where the project involves reusing an existing building, the WBLCA should include only the elements newly brought to site, plus the burdens from any on-site repairs and demolition. Otherwise, this chapter does not cover the reuse of existing buildings. Currently, green building programs offer greater reward for use of existing buildings, and on-site reuse of existing building components, through means other than WBLCA.

BILL OF MATERIALS ADJUSTMENTS

The WBLCA should consider the fact that a greater quantity of salvaged material than virgin material may be necessary to procure in order to provide the finished quality material installed during building construction. In general, reusing framing elements that were originally designed as a "kit of parts" and are repurposed as such, will generate less waste than reusing stick-built elements when incorporating in a new design.

LIMITATIONS

When determining the material quantities for the reference building, the project team should ensure that the building's bay size, programming, and loading are as equivalent as possible to the proposed building.

Deviations from Standard Industry Practice

The team should not take LCA credits for reusing items in the proposed design when the industry standard or contractor preference is already to reuse material (e.g., reusable concrete forming, crushing existing concrete for use as aggregate in site work). WBLCA should include these standard practices in both the reference building and the proposed design.

SPECIAL CONSIDERATIONS

The engineer of record should evaluate salvaged or reclaimed materials to determine their suitability for reuse. This evaluation could take the form of coupon testing, visual inspection, nondestructive testing, or other assessments that are beyond the scope of this document.

MATERIAL-SPECIFIC STRATEGIES

Wood

Salvaged wood items should be wood that has been reclaimed, either in whole form or partial form, from the project's building site or another site or repurposed on the same project.

Example

US Army Corps of Engineers Federal Center South, Seattle, WA: Heavy timber framing members from an existing on-site warehouse were salvaged and repurposed into composite framing members (Swenson and Black 2013).

Steel

Salvaged steel should be steel that has been reclaimed, either in whole form or partial form, from the project's building site or another site. Steel in a fabricator's inventory that was initially allocated to another project but not used is not eligible for salvaged material embodied energy savings.

Example

National Renewable Energy Laboratory project, Golden, CO: A mix of salvaged and reclaimed gas pipeline was reused as main structural columns in this research facility (Acree Guggemos and Franklin 2011).

Precast Concrete

Salvaged precast should be precast concrete elements that have been reclaimed or salvaged from the project's building site, or another site, and incorporated in the proposed building design. More common examples of salvaged precast may include tilt-up walls, retaining walls, barriers, stadium seating elements, and risers.

Example

Tim Horton's (a coffee company) Office and Distribution Center, Calgary, AB. Precast wall panels were removed and reused for an expansion of the building while the space remained in service (McCraven 2017).

Design for Deconstruction

SCOPE

This chapter provides the user with guidance for implementing the strategy of design for deconstruction (DfD), which is to design structural systems for disassembly so that the structural elements may be reused without re-entering a material manufacturing process. Minor mechanical processing, such as cutting, crushing, grinding, and drilling holes through elements, is acceptable for reuse, whereas melting, forming, or making new material is considered recycling. Projects teams that are already using salvaged materials for reuse should refer to the chapter titled "Incorporating Salvaged Materials."

The heavy industrial manufacturing processes involved to create most structural materials for commercial use in the United States typically demand the most energy and generate the greatest burdens on the environment of all the activities in a structure's life cycle. DfD differs from reuse in that the elements in a project employing reuse are already on their second life while the elements in DfD are most likely in their first life. The team may combine both strategies on a single project, in which case reused elements in their second life are designed for deconstruction to be used for a third life.

DfD is a strategy that aims to extend the life of structural elements after their first life when materials were procured through a conventional manufacturing process. Whether the elements will actually be reused inherently entails some uncertainty due to the relatively long life and unpredictable fate of structural elements at the end of the building's life.

As a result, the WBLCA comparison of DfD to traditional design is mostly academic but is as valid as other comparative calculations that serve design-based rating systems. In most green building programs, the project teams seek recognition in energy and water savings using calculations based on their designs, well before actual energy performance and water use can be measured in the functioning building. Like energy modeling results and water use calculations, the results of a WBLCA are not meant to predict actual performance but to guide and incentivize the design team toward a more environmentally beneficial outcome.

BILL OF MATERIALS ADJUSTMENTS

The BOM should not differ between the baseline and proposed systems, except where the team has intentionally selected different elements or connections to enable deconstruction for future reuse.

LIMITATIONS

Proposed Case

Since the objective of DfD is to facilitate the reuse of structural components, the calculation presented in the methodology below incorporates a reduction in the LCA factors based on reusability. A limitation is that the reduced LCA impact factors accounting for reusability may only be applied to components designed for deconstruction in order to enable reuse. For the remaining structural elements, LCA impact factors should be equivalent to those used for the reference building.

Notes

If using the World Steel equation for cold-formed steel, the team should use the World Steel values in the equation for all elements not designed for deconstruction, and only use values relevant to DfD for the elements that were designed for deconstruction. Likewise, for cold-formed steel, the team may use the same formula for structural materials in both the reference and proposed cases but change the values in the formula for recycling or DfD, whichever end-of-life scenario is applicable to the components in the design.

SPECIAL CONSIDERATIONS

Phases

It is a violation of LCA fundamentals for a project team to take credit for benefits within a study boundary that another team has also taken credit for, often referred to as "double-dipping." Specifically regarding DfD, it is not acceptable for a project team to take credit for design for reuse in a future project if that project (or projects if the pieces are spread across multiple buildings) takes any overlapping share of credit. There are a few ways to avoid violating this prescription. One is to only account for the benefits of DfD in the second project(s) but not in the first project. Another option is to allocate some credit to the first project and the remaining credit to future project(s). The last option is to allow only the first project to take credit.

In the first option, the team would use the guidance presented in the "Incorporating Salvaged Materials" chapter only for the project that benefitted from

DfD. In this case, the user would not include Module D (see System Boundary Table 1: Life Cycle Stages and Corresponding LCA Strategies), and more standard procedures could be used where the resulting discount would appear in Module A. In conjunction, the team would not claim credit for the first project, even though it may have helped to enable the reuse of elements for the second project(s).

Notes

This approach is more reflective of current standard practice in LCA but does not incentivize DfD. Within green building programs, incentives can be created outside of requirements or credits for WBLCA. However, most green building programs currently lack incentives for DfD. For this reason, the authors chose to explore the topic of rewarding DfD within WBLCA in this document.

If the team takes one of the other approaches, where the benefits of DfD are credited either in whole or in part to the first building, the WBLCA should include all life cycle phases. It is most appropriate to assign the avoided burden for potential future reuse to Module D: Beyond Building Life. Note, however, that current standards do not require inclusion of Module D, and the WBLCA preparer should report any results associated with Module D separately from the results of Module A through C. Furthermore, if the green building program dictating the WBLCA does not include Module D, the project team may need to request the green building program to accept an alternate compliance path to account for DfD within the assessment.

Notes

Note that ISO 21930 requires any environmental impacts or information outside of Modules A through C to be reported separately under "Additional Environmental Information." In no case should the user combine values from Module D with the impacts of Modules A through C and report them as the complete WBLCA results.

Calculation Methodology

The WBLCA should account for savings in environmental life cycle impacts anticipated from structural DfD in the end-of-life phase of the proposed building model, which may be accomplished through a mass-based discounting methodology. In mass-based discounting, the user starts with the full initial environmental burden assuming business-as-usual conditions, and then subtracts the burdens that have been avoided through the building design. The WBLCA output should present the business-as-usual impacts in the scopes of Modules A through C, and the subtracted burdens in the scope of Module D. In doing so, the user will likely find a positive number representing Modules A through C and a negative number representing Module D. Some sample equations can be found in "Appendix A: Discounting Methodologies."

Notes

Both the equations presented in Appendix A are derived from methodologies created for the steel industry to account for the recyclability of steel products at their end of life. Using these equations for DfD is not common within current LCA practice.

Acceptable Sources for Equation Values

The project team should derive the values used for this equation from the most current and relevant industry reports or experts, preferably third-party reviewed, from within the last five years, and these values should be geographically appropriate and used within LCA tools or industry-wide EPDs. For instance, the values for yield should come from a fabrication shop that has experience with repurposing structural elements for reuse, and the burden of repurposing can be determined through consultation with experts knowledgeable in the manufacturing energy demands of repurposing.

For steel buildings, the values for all elements not designed for deconstruction and all elements of the reference building should be obtained from the North American steel values.

Notes

The North American steel AISI LCI values account for the general recyclability of steel, (note that AISC LCI does not), but do not account for the reusability of certain structural steel products. In the WBLCA, the reference building should assume steel recycling at end-of-life because this is current industry practice. The reference building can also use the global average because recycling and production of structural steel is highly integrated into a global supply chain.

Five years is the typical period of validity of environmental product declarations.

Defining the Reference Case

When demonstrating DfD, the reference design should be equivalent to the proposed building, except for components specifically designed to enable deconstruction for reuse at the end of the building's service life. The structural system of the reference building should maintain the same functional equivalence in all other aspects. It should also be of a similar material to the proposed building. The only exception allowable is if an earlier design option used a more dissimilar material. For this exception, the reference building should be from the same project and should meet the same programmatic requirements.

Notes

For example, a steel building designed for deconstruction in the proposed building should remain a steel building in the reference building, unless project-specific documentation can be supplied that shows a non-steel building was actually considered earlier in design. The reference building may employ unconventional combinations of materials, such as precast concrete over steel-framed systems, when the project team can justify through a reference building option why they are not using an industry norm as the reference structural system.

Supporting Documentation for DfD

The project team should follow the documentation requirements prescribed by the green building program. At present, however, no green building program spells out the documentation requirements for WBLCA using DfD; therefore, this book provides some best practice guidance, as described below.

The project team should provide project-specific documentation to support that the system was designed for deconstruction. This evidence should support the quantities of DfD elements versus the non-DfD elements of the structure used in the chosen equation, which should include the following items at a minimum:

- An inventory of the elements that are technically feasible to deconstruct and reuse, highlighting the ones that would not have been under conventional design,
- A narrative of the design decisions made by the project team to enable deconstruction and reuse of the building elements in the inventory above.

Further evidence can also include the following:

- A DfD plan,
- Deconstruction/disassembly drawings,
- Records of discussions with deconstruction contractors regarding the feasibility of the design for deconstruction,
- Record that a contractor and/or structural engineer with disassembly experience have reviewed the drawings and DfD plan.

Notes

The authors of this guide searched for but could not find team member accreditations or licensures that could distinctly qualify a DfD project. However, training programs exist for contractors to help them learn how to disassemble buildings to minimize damage to materials and maximize their opportunities for reuse.

Participating in Operational Energy Savings

SCOPE

This chapter provides guidance for implementing the strategy of including operational energy savings in the WBLCA. Energy savings can be realized when the project team designs holistically and considers the impacts structural materials have on the operational energy use of a building. The WBLCA can account for these savings if the operational energy phase is included in the assessment. By including this phase, the project team can consider tradeoffs between embodied and operational impacts in the selection of materials and design of structure. Reduction of MEP systems due to changes in the structural design is further discussed under Integrated Design Strategies in the "Nonstructural Material Quantity Reduction" chapter.

BILL OF MATERIALS ADJUSTMENTS

If the WBLCA preparer includes the operational energy use phase in the assessment, the BOM should reflect any impacts on material quantities that occur because of strategies implemented to improve operating energy performance. BOM impact includes any increase in material quantities used for improving operating energy performance even if the BOM adjustment is not required for structural performance.

Notes

Certain material strategies to improve operating energy performance, such as increasing thermal mass, will result in more structural material use. Including the operational energy use phase in the WBLCA can offset increased embodied impacts as a result of increased material use and demonstrate an overall environmental benefit.

LIMITATIONS

If the WBLCA preparer incorporates the operational energy use phase into the assessment, this guide recommends using the Early Design Option for the reference building. The Existing Building and Building Archetype Options for the reference building may be used but are not advised. The designated energy professional on the project should determine operational energy use for both the design and reference buildings through analysis or professional judgment.

Notes

The operating energy use of the reference building needs to be known so that the user can make an appropriate assessment when employing this strategy. The Existing Building and Building Archetype Options for the reference building are less likely to have energy performance data available. It is not appropriate to use an operating energy use database, such as the Commercial Buildings Energy Consumption Survey, to substitute for operating energy use unless that data can be associated with a specific reference building design.

An earlier design will likely have engaged the entire design team, meaning that the designated energy professional is more likely to have knowledge of the energy performance of the reference building or to have already conducted an analysis of the operating energy use.

SPECIAL CONSIDERATIONS

Phases

Not all green building frameworks include the operational energy use phase in the WBLCA. The user is referred to the applicable green building code, standard, or rating system to determine if the operational energy phase should be included.

The WBLCA should include the operating energy use phase for both the reference building and as-built building if it is included in either. The project team should agree on whether to include the operational energy phase when the scope of the WBLCA is defined (refer to the Team Members section in the "General Considerations" chapter for further explanation).

Minimum Energy Performance

The WBLCA should show that the minimum energy performance, as stipulated by the applicable green building code, standard, or rating system, is met for both reference building and as-built building, regardless of whether or not the operating energy use phase is included in the assessment. If the applicable green building framework does not prescribe a building performance metric, then the owner or design team should dictate the minimum performance requirement.

Notes

To best demonstrate the positive effect of material strategies on operational energy use, the WBLCA should utilize a reference building design that meets the minimum energy performance requirements only, as prescribed by the relevant code, standard, or rating system, to achieve the most beneficial results.

Exclusion of Operational Energy Use Phase

If the WBLCA does not consider the operational energy use phase, most green building programs require the operating energy performance of the proposed building to be functionally equivalent to that of the reference building. Functional equivalency can be demonstrated by the professional judgment of the energy professional or through more thorough assessment of the energy performance. The user is referred to the applicable green building program for functional equivalency requirements.

Notes

Demonstrating operating energy functional equivalency prevents the embodied impacts of a building design from being improperly penalized if strategies to reduce operating energy increase material use.

The energy professional might not consider the effects of localized structural detailing impacts, such as thermal bridging, when determining functional equivalency. The WBLCA should include these effects where practical. The WBLCA preparers should consult with project team members who have energy performance expertise (such as the energy modeler, mechanical engineer, enclosure designer, or architect) regarding the effects of structural materials on building energy performance.

REFERENCES

Acree Guggemos, A., and Franklin, S. (2011). "Greening steel construction." *Mod. Steel Const.* https://www.aisc.org/globalassets/modern-steel/archives/2011/05/2011v05_greening_steel.pdf (Accessed Apr. 26, 2018).

Athena guide to WBLCA in green building programs. (2014). http://calculatelca.com/wp-content/uploads/2014/03/Athena_Guide_to_Whole-Building_LCA_in_Green_Building_Programs_March-2014.pdf (Accessed Apr. 26, 2018).

McCraven, S. (2017). "Sustainability and practicality of precast concrete." National Precast Concrete Association http://precast.org/sustainability-and-practicality-of-precast-concrete (Accessed Mar. 1, 2017).

Simonen, K., et al. (2015). "Integrating environmental impacts as another measure of earthquake performance for tall buildings in high seismic zones." *Proc., Structural Engineering Institute Structures Congress*, ASCE, Reston, VA.

Swenson, J., and Black, J. (2013). "A worthy wager." *Structure* http://www.structuremag.org/?p=449 (Accessed Apr. 26, 2018).

APPENDIX A
Discounting Methodologies

This section offers some suggestions for the calculation of environmental impacts in the case of a system that was designed for deconstruction. There are no prescribed formulas or methodologies to use to account for the avoided burdens of deconstruction and reusability of building components, but some methods developed for recyclability are presented following. How to adapt these methods to apply to DfD is an active research topic.

The following equation is offered by Eaton and Amato:

$$S_{ei} = (P_{ei} - R_{ei}) * \frac{(1 - r)}{(1 - r^n)} + R_{ei}$$

where

S_{ei} = embodied impact of material discounted for secondary use (reuse and/or recycling)
P_{ei} = embodied impact of primary (virgin) material
R_{ei} = embodied impact of reused or recycled material
r = rate of reuse or recycling
n = number of times reused or recycled

Another possible equation to use is the mass-based discounting methodology offered by World Steel

$$S = X_{gt} - (RR - s) * Y * (X_{pr} - X_{re})$$

where

S = embodied impact of material discounted for secondary use (reuse and/or recycling)
X_{gt} = embodied carbon of material from cradle to gate
RR = recycling rate
s = scrap
Y = yield (always <1)
X_{pr} = embodied impact of 100% primary (virgin) material
R_{ec} = embodied impact of 100% reused or recycled material

This equation was used by Eaton and Amato in an LCA study for the Steel Construction Institute in 1998 and adapted from the equation found in a paper by Amato, Brimacombe, and Howard (1996). The World Steel Methodology differs slightly from this formula.

The discounting methodology is used by World Steel but only currently accounts for recyclability (not reuse) of steel in the World Steel LCA impact numbers.

REFERENCES

Amato, A., Brimacombe, L., and Howard, N. (1996). "Development of quantitative methodology for assessing embodied energy of recyclable and reusable materials/products." *Ironmaking Steelmaking*, 23(3), 235–241. http://www.lcaforum.ch/Portals/0/DF_Archive/DF33/Thomas%20-%20ArcelorMittal%20-%20LCA%20DF33.pdf (Accessed Mar. 1, 2017).

APPENDIX B
LCA Tools

See the following LCA tool descriptions and table.

Building-specific tools suitable for non-LCA specialists:

- Athena Impact Estimator for Buildings (IE): The Athena IE is a free LCA tool developed in Canada by the Athena Institute for North American buildings. The tool includes city-specific and national-average data. The user may select predefined building assemblies or input a bill-of-materials. The tool includes a wide range of structural, envelope, and finish materials.
- eToolLCD: The etoolLCD was developed by etool as a web-based Life Cycle Design application for the built form and it accounts for impacts over a product's entire life span.
- Tally: Tally is a Revit add-in developed by KT Innovations (an affiliate of KieranTimberlake Architects), thinkstep (formerly PE International), and Autodesk. The tool includes a wide range of building materials with environmental impacts derived from the GaBi LCI database.
- Environmental Analysis Tool (EA Tool): Developed by Skidmore, Owings & Merrill LLP and released in 2013. The LCA tool quantifies the embodied carbon dioxide equivalents incurred over the lifetime of a building and includes an approximation of the amount of those impacts that can be attributed to repairs of seismic damage. This cradle-to-grave life cycle approach can allow designers, contractors, and owners to understand the implications of their projects early in the design process, quantify all the variables associated with embodied carbon dioxide during the construction phase, and measure the probable seismic damages in fiscal and carbon metrics during the service life of the building. The tool does not account for construction processes or operational impacts unless directly input by the user.

General-purpose tools best suited for LCA specialists:

- GaBi: GaBi is an LCA tool developed by thinkstep, an international company based in Germany. LCI data sources are from both the European and North American markets. GaBi is not building-specific, so it includes a very wide range of materials.
- SimaPro: SimaPro is an LCA tool developed by Pré Sustainability, an international company based in the Netherlands. Earthshift distributes SimaPro for the North American market. Like GaBi, SimaPro is not building-specific and includes a very wide range of materials.

LCA Tool Name	Distributor	Cost	Pros	Cons	Website
LCA Tools for Buildings:					
Athena Impact Estimator (IE)	Athena Sustainable Materials Institute	FREE	Specific to US regions. Material LCI reports all available for free online. Documents online state assumptions behind assemblies. New additional BOM module allows import of BIM takeoffs.	Limited number of assemblies.	www.athenasmi.org
eToolLCD	eTool	$(?)	Has gained some popularity in Australia and UK. Easy options comparison. Allows one to enter info from EPDs. Can use at early design phases. Developers also do buildings assessments.	No visual of the building. Hard to tell where US data comes from.	http://etool.net.au/software/
Tally	KirenTimberlake	$	Revit application that takes quantities from 3-D Revit BIM model and runs it through an LCA	Must have Revit model of building	http://choosetally.com/
EA Tool	Skidmore, Owings & Merrill LLP	FREE	Easy to use. Can run an LCA with limited input. Quantifies environmental impacts associated with seismic damage	Not for commercial use.	https://somhpd.com/eatool/
General Purpose LCA Tool (geared for LCA Practitioners):					
SimaPro	PRe	$$	Customizable. Access to wealth of LCI data. LCI's are near current and well maintained.	Packages range from $4,000-$10K. Most building assemblies would need to be created from scratch. Most LCI data is European.	http://www.pre.nl/life_cycle_assessment/
GaBi	(formerly thinkstep/PE International)	$$	Recent add-ons for non-LCA-expert interface.	Similar to SimaPro. Each database module $1,000-$10K	https://www.thinkstep.com/industries/building-construction/

$ = < $1,000 / year or one-time purchase
$$ = > $1,000 / year or one-time purchase

APPENDIX C

Additional Resources and Case Studies

This list is taken mostly from the SEI Sustainability Committee Top 10 LCA FAQs Bibliography found at http://structureandsustainability.blogspot.com/2014/09/top-10-lca-faqs-bibliography.html.

Anderson, J., and Silman, R. (2009). "A life cycle inventory of structural engineering design strategies for greenhouse gas reduction." *Struct. Eng. Int.*, 19(3), 283–288.

Arup. (2010). *Embodied carbon study: Study of commercial office, hospital and school buildings*, The Concrete Centre, London.

Cole, R., and Kernan, P. (1996). "Life-cycle energy use in office buildings." *Bldgs. Environ.* 31(4), 307–317.

Comber, M. V., and Poland, C. (2013). "Disaster resilience and sustainable design: Quantifying the benefits of a holistic design approach." *Proc., Structural Engineering Institute Structures Congress*, ASCE, Reston, VA.

Curran, M. A. (2006). *Life cycle assessment: Principles and practice*, EPA, Cincinnati, OH, 80.

Fernandez, N. P. (2008). *The influence of construction materials on life-cycle energy use and carbon dioxide emissions of medium size commercial buildings*, Victoria University of Wellington, Wellington, NZ.

Hsu, S. (2010). *Life cycle assessment of materials and construction in commercial structures: Variability and limitations*, Massachusetts Institute of Technology, Cambridge, MA.

Kaether, B. (2012). "Embodied CO2 of structural frames." https://www.istructe.org/journal/volumes/volume-90/issues/issue-5/articles/embodied-co2-of-structural-frames (Accessed Apr. 26, 2018).

Kneer, E., and Maclise, L. (2008). "Consideration of building performance in sustainable design: A structural engineer's role." *Proc., Structural Engineers Association of California (SEAOC) Annual Convention.*

Konig, H., Kholer, N., Kreissig, J., and Lutzkendorf, T. (2010). "A life cycle approach to buildings: Principles, calculations, design tools." Radaktion DETAIL, Munich.

KT Innovations. (2016). "Tally LEED v4 LCA guide." Documentation Guide, KT Innovations, Philadelphia. http://choosetally.com/resources/ (Accessed Apr. 26, 2018).

LEED. (2012). *Reference guide for green building design and construction, version 4* https://www.usgbc.org/resources/leed-v4-building-design-and-construction-current-version, U.S. Green Building Council, Washington, DC (Accessed Apr. 26, 2018).

Ochsendorf, J., et al. (2011). *Methods, impacts, and opportunities in the concrete building life cycle*, Massachusetts Institute of Technology, Cambridge, MA.

Preservation Green Lab. (2011). *The greenest building: Quantifying the environmental value of building reuse. National Trust for Historic Preservation.* http://www.preservationnation

.org/information-center/sustainable-communities/green-lab/lca/The_Greenest _Building_lowres.pdf (Accessed Jul. 11, 2014).

Ramesh, T., Prakash, R., and Shukla, K. K. (2010). "Life cycle energy analysis of buildings: An overview." *Energ. Buildings*, 42, 1592–1600.

Simonen, K. (2014). *Life cycle assessment*, Routledge, New York.

Simonen, K., et al. (2015). "Integrating environmental impacts as another measure of earthquake performance for tall buildings in high seismic zones." *Proc., Structural Engineering Institute Structures Congress*, ASCE, Reston, VA.

Webster, M., Kestner, D., Parker, J., and Waltham, M. (2007). "Deconstructable and reusable composite slab." Lifecycle Building Challenge: Winners in the building category: Component—Professional Unbuilt http://www.lifecyclebuilding.org/2007.php (Accessed Apr. 26, 2018).

Index

Page numbers followed by *f* and *t* indicate figures and tables, respectively.